NUMBERPOWER

in Nature, Art, and Everyday Life

Keith Ellis

ST. MARTIN'S PRESS   NEW YORK

Copyright © 1978 by Keith Ellis
All rights reserved. For information, write:
St. Martin's Press, Inc., 175 Fifth Ave., New York, N.Y. 10010
Printed in Great Britain
Library of Congress Catalog Card Number: 78-3124
First published in the United States of America in 1978

Library of Congress Cataloging in Publication Data

Ellis, Keith.
    Numberpower: in nature, art, and everyday life.
    Published in 1977, in England, under same title.
    Bibliography: p.
    Includes index.
        1. Numbers, Theory of.   2. Cycles.   3. Symbolism of
numbers.   I. Title.
QA241.E44   1978   133.3'35   78-3124
ISBN 0-312-57988-8

CONTENTS

38719

For Myfanwy Rose

ACKNOWLEDGMENTS

The author wishes to thank the following for their kind permission to reproduce copyright material:

Faber and Faber Limited for the quotation on page ix and Figure 12 on page 167 from *The Modulor* (1954) by Le Corbusier; also for Figure 13 on page 169 from *Modulor 2* (1958) by Le Corbusier.

The Hutchinson Publishing Group Limited for Figures 1 and 2 on page xii from *Animal Architecture* (1975) by Karl von Frisch; also for Figure 1 on page 146 from *The Anticipation of Nature* (1965) by Rom Harré.

McGraw-Hill Book Company (UK) Limited for Figure 1 on page 6 from *Number Theory and its History* (1948) by Oystein Ore.

Mr N. J. Chipping, L.D.S., R.C.S.Eng., director of the Biorhythmic Research Association, for Figure 1 on page 42 and Figure 2 on page 43.

The Ladbroke Casino Division for Figure 1 on page 77, Figure 2 on page 79, Figure 3 on page 80 and Figure 4 on page 88.

The *Financial Times* and *Investors Chronicle* for Figures 1 and 2 on page 106.

Heinemann Educational Books Limited for Figure 2 on page 149. Academy Editions for Figure 7 on page 158, Figure 8 on page 159, and Figure 9 on page 161, all from *Architectural Principles in the Age of Humanism* (Academy Editions, London) by Rudolf Wittkower.

Harvard University Press and William Heinemann Limited for Figure 3 on page 179 from Plato's *Timaeus*, Loeb Classical Library (1929) translated by R. G. Bury; also for Figure 1 on page 198 from *The Copernican Revolution* (© Harvard University Press, 1970) by Thomas S. Kuhn.

'The chamois making a gigantic leap from rock to rock and alighting, with its full weight, on hooves supported by an ankle two centimetres in diameter: that is challenge and that is mathematics. The mathematical phenomenon always develops out of simple arithmetic, so useful in everyday life, out of numbers, those weapons of the gods: the gods are there, behind the wall, at play with numbers.'

Le Corbusier[1]

[1] See bibliographical notes beginning page 217

Introduction

IF you have the good luck to know a beekeeper, ask to see one of his honeycombs. The cells are built with the precision of a watchmaker. When construction starts, the bees gather into a close mass which keeps the temperature steady between 34 and 35 degrees centigrade and enables them to secrete wax under their bellies, mix it with saliva and add it, still malleable, to the comb they are working on.[1]

As each new layer of wax is added, the bees probe the cell walls with tiny spikes on the ends of their feelers. They press down the surface with their jaws and allow it to rebound. The thickness of wax built up determines the speed of the rebound which they register through organs of touch on their feelers. They then know whether a little more wax needs to be added or a surplus scraped off. In this way, they build the walls of worker cells to a uniform thickness of 0·073 millimetres with a latitude of only 0·002 millimetres either way. Drone cells are 0·021 millimetres thicker.[2]

Building proceeds in several places at the same time. Individual bees stay on the job for an average of less than a minute but the various sections eventually join up so exactly that it is impossible to see the joins. Each worker cell measures 5·2 millimetres from wall to wall. Drone cells are a millimetre wider. All slope backwards and downwards at an angle of 13 degrees to the horizontal, just enough to stop the honey running out.[3]

The most elegant aspect of the honeycomb is the shape of the cells. They are perfect hexagons joined together by common walls to form a continuous matrix. All honey-bees build hexagonal cells and if the beekeeper finds it convenient, as many do, to give them a start with their annual comb-building, he has to use a hexagonal framework. It is the only kind his bees will tolerate.

Why hexagons? Why not cells with 3, 4, 5, 7, or 8 sides? Why not circular cells?

The Austrian zoologist Karl von Frisch has given us the answer. If the comb consisted of circular cells, there would be intervening gaps. Space would be wasted and so would wax, for each wall would enclose only one single cell. It could never serve as part of a common wall for an adjoining cell (Figure 1). Cells with 5, 7, or 8 sides could not form a continuous matrix either and again the gaps would waste both space and wax. Triangles and squares can be made to form a continuous matrix with common walls and without gaps. So it might seem that these shapes would be equally suitable. However, a quick calculation shows that the lines of a matrix covering a given area are a good deal shorter if hexagons are used. So hexagons are the most economical form and are used by all honey-bees.[4] (Figure 2.)

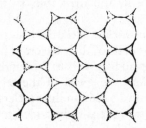

FIGURE 1

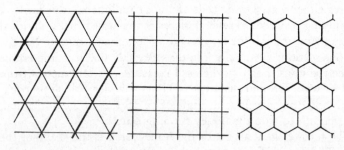

FIGURE 2

Number patterns the whole of our universe, just as it patterns the honeycomb. If it doesn't force itself on us, we instinctively seek it. For we are a part of the universe. We carry the same number patterns within us and we are seeking our own. When we find them we feel a thrill of recognition. How else explain our delight at the geometric intricacies of Islamic art or the rose window of a Gothic cathedral?

All our lives we are testing these inbuilt number patterns against the raw material of experience. When they fit, we make a discovery that may be sufficient unto itself or act as a sign-post to further discoveries. As the philosopher Auguste Comte remarked, 'There is no enquiry which is not finally reducible to a question of numbers.'[5]

As a tool, number is capable of infinite refinement. Our usual notational system consists of only ten symbols but with the help of computers, we have elaborated arithmetic to lengths unimaginable fifty years ago. We now know of a prime number with 6,002 digits ($2^{19937} - 1$) and mathematicians have calculated the square root of 2 to 1,000,082 decimal places. In 1973, the French mathematicians Jean Guilloud and Martine Bouyer published a 200-page book whose text consisted of a single number, the ratio of the circumference of a circle to its diameter (π) calculated to 1 million decimal places. It starts 3·141592653589793 . . . and ends . . . 5779458151 with 999,975 digits in between.[6]

By counting and calculating, ordering and relating, generations of scientists have used number to build the towering edifice of systematic knowledge that we now possess. Through number, they have been able to record and communicate almost every kind of observation and natural law, making it possible for man to achieve the transition from nomad to astronaut in only 10,000 years, a mere eyebrow twitch on the cosmic time scale.

Number has also enabled us to delve into the remotest recesses of the earth's past, and predict, sometimes, its future. We can plot our changing position in space with an uncanny degree of accuracy. In 1974, Dr Thomas van Flandern of the U.S. Naval Observatory in Washington, D.C. noticed from records of the previous twenty years that the moon was gradually slowing down in its orbit round the earth. The most likely reason was a dwindling in the force of gravity, which meant that earth and moon were gradually drifting apart. He discovered that the distance between them, some 238,800 miles, was increasing by an inch a year. None of this could have been worked out without number. The discrepancy would never have been detected without the atomic clocks which now measure time to an accuracy of one second in 60,000 years.[7]

Also in 1974, two teams of physicists working independently at Stanford University and the Brookhaven National Laboratory in America both discovered a new sub-atomic particle to add to the 200 already known. They were already familiar with 'negative mass', which means that things rush towards us when we push them away, and also with the seemingly unscientific word 'strangeness' used to describe other unexpected properties. The new particle was different again. It bore a charmed life and the term 'charm' was added to strangeness, electric charge and atomic number as a factor influencing the behaviour of such particles. Again, charm could never have been discovered without the precise measurement of time made possible by the use of number. Other particles broke down after an existence of only a million-million-million-millionth of a second. The new particle survived between a thousand and a million times longer than this.[8]

Number never diminishes. It gives further significance to something we already know. Whatever meaning we find in a sonnet by Keats or Wordsworth is enhanced by the strictness of its 14-line form just as a musical idea is amplified by, say, the sonata form, or a theme in painting or sculpture by use of the Golden Mean. In each case, the medium adds to the message.

Sometimes, too, number introduces an extra dimension. All individual numbers are symbols standing for the abstractions of oneness, twoness, and so on. When a particular abstraction is also held to represent some truth about the universe, its incorporation in a work of art gives added meaning. In Christian art, the Trinity may be symbolized by a clover leaf or three interlocking circles, the Cross by a knot with four loops or by four leaves arranged symmetrically. In Byzantium, the number 6 was thought to reflect the universal harmony because it was 'perfect', i.e. the sum of its own factors $(1 + 2 + 3)$. In Byzantine mosaics, we often find flowers with six petals or sheep grouped in sixes.

The feeling that some numbers have a significance beyond themselves affects us too in our everyday lives. We read of families who insist on having the number of their home changed from 13 to 12a because they believe that 13 is unlucky. We may think them superstitious but they are no more so than the tycoon who pays a large premium for a car with a single-digit registration number. To him, the single digit confers a status denied to owners of cars with the more usual multi-digit numbers.

In the stock market, brokers recognize that some numbers influence buying decisions. At the beginning of 1976, the *Financial Times* ordinary share index persisted in hovering around the 400 mark. Investors tended to hold back when it rose above 400 because shares were then thought to be expensive. As soon as it slipped below 400, they were thought to be cheap and a spate of buying quickly chased it back. There was no logical reason for regarding 400 as the 'right' level. It could just as well have been 390 or 410. But in the minds of investors, 400 took on a power beyond rational explanation because it was a round number.

Significantly, the other factor that was holding back the *FT* index in Britain was the Dow Jones index in America. British investors, like those in many other countries, were hesitating until it had convincingly broken through the even more magical level of 1000.

It seems inescapable, then, that numbers have *in themselves* a power over us that can be demonstrated by concrete evidence. We may say that it is subjective or irrational but it exists. We can point to innumerable cases where fear of 13 or a liking for single digits or round numbers has cost people money, a criterion as objective as even the most sceptical could wish for.

Could numbers also have a power that emanates from a source outside ourselves? It is a fact of history that every American president elected at intervals of twenty years since 1840 has died in office. Does the number 20 exert some baleful influence over the American presidency or is the continuing fate of the twenty-year presidents just a coincidence? If so, what precisely does 'coincidence' mean? In the last three hundred years, we have tended to think that events can be linked only by cause and effect. Could there be some other connecting principle that we have lost sight of?

These are questions that challenge the usual boundaries of scientific knowledge. They can no longer be ignored. In this book, I want to look at some of the ways in which the power of number is made manifest, in art as well as in arithmetic, in living myth as well as in the number-based technology that took men to the moon. All I ask of the reader is an open mind.

I

To the Power of Ten

WHY do South African tree crickets prefer to sing from a hole in a sun-flower leaf?

Like other crickets, they produce their chirps by scraping a file-like set of teeth on one front wing against the sharp leading edge of the other. This makes both wing surfaces vibrate, creating a 'song' that other crickets can pick up by means of hearing organs on their front legs. Most of the chirping is done by males either to scare off rivals or to attract or stimulate females. So the pitch and rhythm of the song varies not just from species to species but also according to its purpose.

Entomologists studying the South African tree cricket found yet another variation. It spends much of its life hopping around sun flowers. When it dug a pear-shaped hole in the centre of a leaf, sat inside with folded legs and sang from there, its chirp was three times as loud as when it sang from the leaf's surface. But why? And what was so special about the sun-flower leaf? What reason could they have for bothering to dig out a hole when there were plenty of ready-made holes in walls or trees?

The answer lay in modern acoustic theory. When the entomologists measured the wavelength of the tree cricket's chirp, they found it was 170 millimetres. They measured the diameter of a typical sun-flower leaf and found it roughly half as wide. We know that the most efficient way of amplifying a sound is to use a baffle whose diameter is about half its wavelength. When the tree cricket sang from its hole, the leaf acted as a baffle, amplifying its chirp to the maximum possible extent. It was the equivalent of a thousand-watt public address system for sending out a mating call.[1]

Number enabled entomologists to plumb the secrets of the

cricket's sex life. Number, too, has enabled us to discover equally precise methods of adaptation in other animals. Some night-flying moths, for instance, have hearing organs at each side of their thorax and it is possible to measure electrically the strength of the messages buzzing along the nerves that lead from them. These messages tell the moth how to react to whatever sound is picked up. Now human ears can normally detect sounds only up to a frequency of 20,000 cycles per second but the moths' hearing organs were found to be specially sensitive to those with frequencies between 15,000 and 60,000 cycles per second. Again, it had to be asked, 'Why?' On the face of it, there was no particular reason why moths should need such a high and wide hearing range.

The answer turned out to be bats. These navigate by sonar or echolocation. As they flit about in their nightly hunt for moths and other prey, they give out up to three hundred pulses a second, and at close range make more noise than a pneumatic road drill. These pulses bounce back off obstacles in the bats' path and by picking up their echoes, they can work out instantly the type and distance of whatever object is reflecting them.

Now the pulses given out by the common bat have a frequency range between 15,000 and 60,000 cycles a second. This is lucky for humans since it puts most of them outside the human hearing range and stops them deafening us. But the moths are automatically tuned in. As soon as they pick up the signal, warning messages shrill along their auditory nerves. Some species simply stop flying and crash-land in a predetermined curve. Bats find this fairly easy to follow and gobble up about half the moths before they hit the ground. The behaviour of other species is more refined. Almost instantaneously, they assess the attacking bat's position by comparing the relative strengths of its signal as received in their left and right ears respectively. In this way, they can tell which side the bat is on. 'Above or below?' and 'ahead or behind?' are decided by the ways in which the signal is modified by the beating of the moths' wings which join their bodies over and in front of their ears. It then takes less than a hundredth of a second for a moth to start evasive action. Some species even have their own transmitter, tuned to the bats' frequency, which has the effect of scaring the bat away when it is switched on by the moth's alerted brain. So by developing ears able to pick up the high frequency signals sent out by one of their main enemies, these moths have greatly improved their chances of survival.[2]

Obviously, we must not make the mistake of thinking that moths consciously, or even unconsciously, worked out this pattern of behaviour, still less that they used number to do so. The matching of their hearing range with the frequency of the bats' signals evolved over millions of years by a process of trial and error, because error meant death and only the tuned-in moths survived to pass on their inbuilt defence mechanism to their off-spring. Number may have enabled us to discover how this defence system works and how it fits in to the stupendous mosaic of number correspondences that pattern the universe. The moths themselves are totally unaware of number.

Discovering number sense in animals is not easy. We come across clues in various quarters but we have to look a long way before we find anything remotely resembling the number sense in man.

Solitary wasps seem to offer a parallel. They lay their eggs in separate cells which they stock with caterpillars as living larders for their unborn young. Females of each species tend to leave much the same number of caterpillars in each cell, some species averaging five, others as many as twenty-four. Females of the genus *Eumenes* even distinguish between male and female eggs, providing some five caterpillars for each male and ten for each female.[3] Then again, some birds set a definite limit to the number of eggs they lay, even though they are physiologically capable of producing more. If a willow warbler's eggs are systematically removed as she lays them, she will rarely continue beyond her usual maximum of six before abandoning her nest and starting all over again with a second clutch.[4] It seems, though, that neither the *Eumenes* wasp nor the willow warbler has any choice in the matter. Their behaviour is controlled by their genes. They no more understand number than a crow appreciates the iridescent sheen of his glossy black plumage.

Even so, there is some evidence that animals can be taught a rudimentary form of counting. As long ago as 1943, the German biologist Lögler trained a grey parrot called Jako to feed himself by numbers. If Lögler flashed a light five times, Jako would gobble up five pellets scattered on a row of trays. After three flashes, he would take only three—and so on. When Lögler substituted notes on a flute for the flashes of light, Jako was just as accurate. Nor did changes in regularity or pitch affect his skill. He even learned how to open boxes with one, or two spots on their lids after one, or two notes on the flute. More recently, it has been shown that a civet cat can be trained to distinguish between odd and even numbers arranged as patterns of spots or other shapes.[5]

Man has developed this ability to a much finer degree. He can abstract. If we are asked what two flowers, two shoes, two microbes, and two spacecraft have in common, we have no difficulty in separating out the idea of twoness.

We can also perceive small numbers directly. In the 1920s, several American psychologists showed fourteen people a random succession of white cards. They were marked with anything from two to fifteen haphazardly arranged black dots. Each card was flashed for only one-tenth of a second but almost all the people tested made 98 or more correct calls out of 100 on cards showing two, three, or four dots. Even with five dots, all but 2 scored 94 or more. It was only when more than eight dots were shown that the overall success rate dropped below fifty per cent. It seems then that the average adult can directly perceive any number up to 8, though there is some evidence that after 5 we may 'cheat' by grouping or counting the dots on a mental image of the card.[6]

Man probably began to use number when he first settled down to farm and trade after the last ice age some ten thousand years ago.[7] His ability to perceive number directly was probably less well developed than ours but it enabled him to discover *without counting* how many cattle or bundles of wheat he owned. He did so by comparing one group directly with another. The vocabulary of some primitive languages makes this clear, since the word for 2 is clearly linked with that for, say, a bird (two wings) and that for 3 with clover (three leaves).[8]

Even now, we sometimes assess numbers by the same method. When we see that all the compartments of a standard half-dozen egg box are full, we know instantly that there are six eggs. We have compared the group of compartments with the group of eggs and found them equal. Similarly, we can even keep a record of numbers without counting by forming a comparable group of pebbles, tying the same number of knots in a piece of string or cutting the same number of notches on a stick.

These methods are not always convenient, of course, as the British government decided after keeping its Exchequer accounts on elm splints for seven centuries. In 1826, it took the revolutionary step of changing to quill pens and ledgers, only to bark its shins on a still knottier problem. What to do with the mountains of notched sticks through which armies of woodworm were eating their way? After eight years' deliberation, the order was given to burn them, whereupon an overheated stove set fire to some panelling and both Houses of Parliament were razed to the ground. To this accident, we owe the present building by Sir Charles Barry and a masterly painting by Turner who stayed up all night to sketch the holocaust.

Apart from the danger of being used deliberately or accidentally as firewood, notched sticks had another drawback. There was a limit to the number of notches that could be carved on them. Heaps of pebbles became inconveniently large and knotted string too long for complicated transactions. It is not surprising, therefore, that man hit on the idea of counting. It was counting that gave us our number system and made possible the study of so many branches of mathematics.

Counting consists of arranging directly perceived numbers in ascending order and giving them names—one, two, three, and so on. When we want to find the size of a group, we check off the individual items against these names, starting with 'one'. The name corresponding to the last item then gives us the number in the group.

From the start, fingers were an obvious aid and we still refer to our ten numerals from o to 9 as 'digits', from the Latin *digitus*, meaning a finger. Most western teachers frown on finger-counting but the Venerable Bede (A.D. 673–735) described a system that went as high as 1,000,000 and as late as the sixteenth century standard texts described systems to 9,000. By then, finger-counting had become surprisingly sophisticated, as we can see from the illustration contained in a Venetian text-book of 1523 (Figure 1). New Guinea tribes still use similar techniques, and within living memory peasants in the Auvergne region of France had a way of multiplying together any numbers between 5 and 10 on their fingers.

They subtracted 5 from each number and turned down in each hand a number of fingers corresponding to the two remainders. The sum of the fingers turned down gave the first digit of the

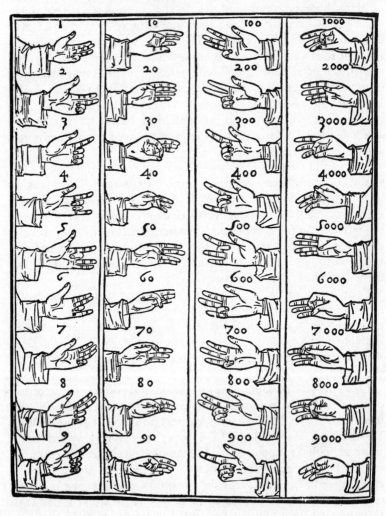

FIGURE 1 Finger numbers

answer and the product of the fingers staying up gave the second.

e.g. To multiply 7 by 8.

7 − 5 = 2. Turn down 2 fingers of the left hand.

8 − 5 = 3. Turn down 3 fingers of the right hand.

Add the turned-down fingers. 2 + 3 = 5 (the first digit of the answer).

Multiply together the fingers remaining up. 3 × 2 = 6 (the second digit of the answer).

Answer = 56.

The method works well, though sometimes the product of the fingers remaining up exceeds 10 and a ten has to be added to the tens column of the answer. According to Professor Tobias Dantzig of the University of Maryland, 'Artifices of the same nature have been observed in widely separated places, such as Bessarabia, Serbia, and Syria. Their striking similarity and the fact that these countries were all at one time parts of the great Roman Empire, lead one to suspect the Roman origin of these devices. Yet, it may be maintained with equal plausibility that these methods evolved independently, similar conditions bringing about similar results.'[9]

However that may be, there was clearly too a need to write numbers down to provide a more permanent record and in a sense devices such as the notched stick fulfilled this need. After all, notches differ little from the I, II, III, IIII of the Roman system which came much later. For large numbers, however, marks of this kind would be inconvenient because there would be too many of them. Long before the invention of written number systems, men hit on the idea of lumping individual items into equal-sized groups and then numbering the groups. When the number of groups equalled the number of individual items in each, the groups were themselves lumped together in a yet larger group, and so on. The number at which the change is made from one tier to the next is called the base. In our own system, the base is 10 but many early systems had a different one.

An incident reported by the Greek historian Herodotus shows how the idea of a base may have been arrived at. He describes the method used by Xerxes, king of Persia, to number his army just before his invasion of Greece in 480 B.C.

'A body of ten thousand men was brought to a certain place and the men were made to stand as close together as possible. A circle was then drawn round them and the men were let go. Then, where the circle had been, a fence was built about the height of a man's waist and the enclosure was filled continually with fresh troops, till the whole army had in this way been numbered.'[10]

A base of ten thousand might have been convenient for counting an army but it was far too large for everyday purposes. It would have needed 10,000 separate digits and multiplication tables up to 10,000 × 10,000. Besides, it seems likely that some primitive tribes were unable to conceive of larger groups than 2, for instance, Australian aborigines of the River Murray region, whose number system went enea — 1, petcheval — 2, petcheval enea — 3, and petcheval petcheval — 4.[11] Even today we use binary, which has a base of 2, for programming computers, so that $1 \equiv 1$, $10 \equiv 2$, $11 \equiv 3$, $100 \equiv 4$, $101 \equiv 5$, and so on.

According to the French mathematician Pierre Laplace, the philosopher Leibnitz saw the binary system as a symbol of creation. He wrote 'He imagined that Unity represented God, and Zero the void; that the Supreme Being drew all beings from the void, just as unity and zero express all numbers in his system of numeration.'[12]

Yet a system consisting of only two symbols is bound to be unwieldy in expressing large numbers. Even 87 becomes 1010111 in binary and many of the numbers used in astronomy would run into several lines of print with a high risk of error. For most purposes, a larger base is more convenient.

Probably the largest in regular use was one of the earliest, that of the Sumerians and Babylonians which dated back to the third millennium B.C. The people of the city states scattered across Mesopotamia needed a system for measuring land, weighing goods, assessing taxes, planning irrigation schemes, and making complicated astronomical calculations, and we still have some of the clay tablets on which they kept records and set out problems in algebra and geometry as school exercises. All are in the sexagesimal system which had a base of 60. Moreover, it was a *positional* system. Each digit represented different powers of sixty according to its position in the number as a whole. Using our own numerals with the Babylonian system, 9, 5, 3 would be equivalent to $(9 \times 60^2) + (5 \times 60^1) + (3 \times 60^0) = 32,400 + 300 + 3 = 32,703$ in our numbers.

The usefulness of this positional arrangement enabled the Babylonian system to survive through Greek and Roman times and indeed, we still use it for measuring the number of seconds in a minute and minutes in an hour. We also use it for scoring at tennis. Originally, scores were kept on a clock-like dial with a movable hand. When a player scored his first point, the hand was moved to a position corresponding to fifteen minutes past the hour, and for his second point to the half-hour or thirty-minute position. So we got '15' and '30'. 'Forty', the score for the third point, is probably short for 45. 'Love' may be a corruption of the French *l'oeuf* (egg), which looks a bit like the zero on a score board. 'Deuce' is a corruption of *deux* (two) and means that the game can only be won when one player scores two points in succession.

As a base, 60 has the advantage of being divisible by 2, 3, 4, 5, 6, 10, 12, 15, 20, and 30, enabling weights, measures, and other units to be sub-divided in many different ways. Its main drawback is the difficulty of memorising multiplication tables that go all the way to 60×60. Despite its obvious attractions, a smaller base was needed.

Logically, a base of 12 might have been best. It is neither too large nor too small and can be divided by 2, 3, 4, and 6. The French naturalist Georges Buffon (1707–88) was the first of many reformers who have advocated a duodecimal base right until the present day. Twelve has a natural appeal. The custom of counting eggs by the dozen shows every sign of surviving metrication in Britain and food manufacturers have stated that they will continue to pack cans in cases of twelve because the 3×4 shape is easier to handle than a metric 2×5.

Moreover, the foot, based on the length of an average man's foot, has been a constant unit of measurement from classical times. Both in Rome and in modern Britain and America, it was divided into twelve inches, each of which was originally equivalent to the width of an average man's thumb. When Britain first decided to go metric, government architects were surprised to learn from their French counterparts that the standard module in which metric buildings were designed was an awkward 300 millimetres. When

they asked why, they were told it was the nearest convenient
equivalent to the English foot, which was considered to be the
most suitable.

Whatever its merits, 12 was foredoomed as a base for a simple
anatomical reason. As we have seen, men have always used their
fingers and, sometimes, their toes too for counting and these did
not fit in with a system based on 12. At first, some peoples probably counted on the left hand only, using the forefinger of the
right as a pointer. This gave them a base of five. A palaeolithic
tally stick found in Moravia in 1937 had 25 of its 55 notches
marked off in sets of 5 and W.C. Eels, who investigated the
number systems of North American Indians, found that many
tribes also used 5.[13] So too, perhaps, did the forerunners of the
Romans, for while V = 5 in the Roman system, 10 is represented
by two Vs, one on top of the other inverted ($\overset{\vee}{\wedge}$ or X), 15 by three
Vs ($\overset{\vee\,\vee}{\wedge}$ or XV) and 20 by four Vs ($\overset{\vee\;\vee}{\wedge\,\wedge}$ or XX). There are, too,
separate symbols for 50 (L) and 500 (D).

Another alternative was to use both fingers and toes for counting and in fact 20 was used much more widely as a base than 5. One
of its advantages was that it had more divisors. Some tribes of
North American Indians preferred 20 and so did the Aztecs of
Mexico who even had a day of 20 hours and Army divisions of
20 × 20 × 20 = 8,000.[14] The Mayas of Central America had a similar
system but with one variation. They were obsessed by the calendar,
and since they divided their year into eighteen months of twenty
days (plus five extra days), they did not square 20 for the second
step. They used a base of 18 giving 360. Subsequent steps rose by
powers of 20. We find indications of a system based on 20 in the
Biblical phrase 'three score years and ten' and even today cabbage
plants are commonly sold by the score in Britain. The traces of a
system based on 20 can also be seen in some French numbers,
such as *quatre-vingt* = 80.[15]

However, finger-and-toe counting was impossible for men
wearing shoes and the alternative of learning multiplication tables
up to 20 × 20 was too irksome. Toes were forgotten and fingers
alone used for counting. Ten quickly became the most widely
used base of all. It was accepted in India by 500 B.C., possibly

earlier in China and in the fourth millennium B.C. in Egypt.
When the Greeks fathered modern mathematics around the sixth
century B.C., they too worked to a base of 10 and it has never had
a serious rival from that day to this.

In itself, a convenient base did not make calculation easy. Many
ancient systems were similar to the Roman in which each higher
step was marked by a different symbol, X for 10, C for 100, M for
1000. Multiples of each step were usually formed by repeating the
same symbol, XX for 20, CC for 200, and MM for 2000. Larger
numbers consisted of unwieldy combinations which had to be
worked out step by step. On one Roman monument, the symbol
for 100,000 occurs 31 times to represent 3,100,000 [16] and even 383
needs ten symbols (CCCLXXXIII). These are clumsy enough but
the real trouble starts when we use them for calculation. How do
you add CCLV (255) to DXXXIII (533)? Worse, how do you
multiply them?

One solution used from early Egyptian times is the method
known as mediation and duplication. It consists of repeatedly
halving the first of the numbers to be multiplied, ignoring
remainders, until it becomes 1. Meanwhile, the other number is
repeatedly doubled in a matching column. When a number in the
first column is even, the corresponding number in the second
column is struck out. The answer is obtained by adding together
the remaining numbers in the second column.

e.g. To multiply 69 by 47.

69	47
34 (remainder 1 ignored)	9̶4̶ (struck out because 34 is even)
17	188
8 (remainder 1 ignored)	3̶7̶6̶ (struck out because 8 is even)
4	7̶5̶2̶ (struck out because 4 is even)
2	1̶5̶0̶4̶ (struck out because 2 is even)
1	3008

Answer = 3243 (= 47 + 188 + 3008).[17]

A more versatile means of calculating was the abacus, a simple device rather like a nursery bead-frame widely used in ancient Egypt, India, Greece, Rome, and mediaeval Europe. Originally, it was probably a simple board on which sand was scattered, columns drawn with a finger to represent successive powers of ten and pebbles placed in the columns to represent the number required, viz:

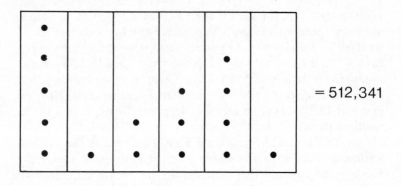

= 512,341

FIGURE 2

Later, round or cone-shaped beads were mounted on rods or wires set in a frame. Either way, it was easy to add two or more numbers, placing an extra pebble, or bead, in the column representing the next higher power of 10 whenever ten had accumulated in a column representing a lower power. Similarly, a pebble could be 'borrowed' from a higher column when needed for subtraction. Multiplications and divisions could be done by successive additions or subtractions.

The abacus was a working model of a simple, efficient system of numerals. It showed that three things were needed to achieve this in a written form. (a) A convenient base such as 10. (b) A *positional* arrangement so that a single set of symbols could refer to different powers of the base simply by their place on the line. (c) A device for making clear in writing the difference between the following arrangements on the abacus:

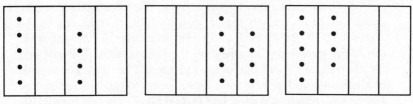

FIGURE 3

All these could be written down as '5 4'. The Egyptians, Greeks, and Romans already had (a); the Sumerians had (b) and (c); the Hindus had (a) and (b). In the event, it was the Hindus who not later than the sixth century A.D. completed the trio by using the zero to indicate an empty column. The three arrangements of the abacus (Figure 3) could now be distinguished clearly as 5040, 54 and 5400.

Indeed, the abacus was no longer needed. The symbols themselves, arranged in their proper positions, could be manipulated more easily than beads. The Hindus found that they could multiply any numbers they wished by drawing a chequer-board of squares, each of which was divided into two triangles by a diagonal. The digits of the first number were placed over the squares, those of the second alongside, so that each square was covered by two separate digits, rather like a map reference. Each pair of digits was multiplied together and the product entered in the appropriate square, the tens in the upper triangle, the units in the lower. The answer was found by adding together the digits in the diagonal columns.

e.g. To multiply 6214 by 123

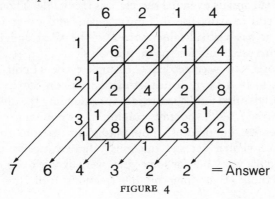

FIGURE 4

How the Hindu system spread to the West is still disputed by scholars. The most likely explanation is that it was taken up by the Arabs who in the seventh and eighth centuries A.D. conquered territories stretching from beyond the Indus, through Persia and North Africa, to Spain and Southern France. They had no numerals of their own and at first used those found locally, including the Hindu system which traders had brought to Persia and Egypt. We know too that the *Siddhantas*, an important Hindu work on mathematics, was translated into Arabic around 773 A.D. Soon, the Hindu system was used throughout the Arab world, though for some reason not generally agreed, the numerals developed in two separate forms. Those now used in the Arab world are derived from the Eastern form and our own numerals, which we call 'Arabic', even though they are Hindu in origin, are derived from a Western form first found in Arab Spain. They were brought to non-Arab Europe by Leonardo of Pisa, otherwise known as Fibonacci, who published his *Liber Abaci* in Pisa in 1202.

Like so many new devices, the Hindu–Arabic numerals were stoutly resisted by the orthodox who stuck to Roman numerals for written work and the abacus for calculation. In Florence, merchants were forbidden to use the Hindu–Arabic system in case their customers should be decieved, but the *abacists*, as they were called, were fighting a losing battle against the *algorists* who followed Fibonacci's lead. The abacus persisted in Spain and Italy until the fifteenth century and in England and Germany until the seventeenth, but the Hindu–Arabic system was obviously superior to the Roman and its eventual success was inevitable. Though the abacus is still in daily use in China, Japan, and other Eastern countries, it has virtually disappeared in the West and Roman numerals are used only for special purposes.

One more step was needed to complete the Hindu–Arabic system. In 1585, a Flemish mathematician, Simon Stevin (1548–1620) of Bruges first published an idea for extending the digits of a number to the *right* of the units column to give *minus* powers of ten, i.e. tenths, hundredths, thousandths, and so on. In other words, he invented decimal fractions. Mathematicians quickly appreciated his work but for some years were unable to agree on a simple method of marking off the fractions from the whole

numbers. For $15\frac{17}{100}$, some wrote 1501172, others ⓪①② or even
1517

⓪①⑪. Eventually, John Napier (1550–1617) invented the
1 5 1 7
decimal point which caught on almost immediately and has been
with us ever since.

Man, then, is the only animal to have a fully developed number
sense. To apply it effectively, we have devised a system of notation
which requires only ten basic symbols. Yet we must not be too
confident. Our mathematical skill has enabled us to leap literally
over the moon but until recently, we have virtually ignored the
power that number exerts over our own bodies. Number may
seem impersonal but almost everything we do is controlled by its
rhythms.

2

What Time is our Body?

NUMBERS can often help us to predict the ebbs and flows in our own lives. They can help us to work out the hours, days, months or years when we are most likely to be alert, vigorous, sluggish, sick, accident prone or jumping with libido. Men have known from the earliest times of the more obvious rhythms that govern our sleeping, waking, eating, and defecating. We now know that life is governed by far more of these rhythms than our ancestors suspected and it seems likely that we are on the brink of discovering others of which we have as yet scarcely an inkling.

These rhythms pulse through every living creature with a precision and subtlety which would often be difficult to detect without the number systems we impose upon time.

We can identify some of them simply by counting the number of events that take place in a given period: midges flap their wings a thousand times a second; canaries' hearts beat a thousand times a minute; lug worms feed ten times an hour. If we watch earthworms nosing their way through a maze, we find that their learning ability reaches a peak once every twenty-four hours during darkness. Activities of this kind are usually regulated by a timing mechanism built in at birth. The Idaho potato, which varies its oxygen intake regularly over a twenty-four hour period, apparently has a separate 'clock' in every individual cell though that of more complex creatures is often localized. The lug worm's is probably in its oesophagus, the cockroach's a factor in its blood. The site (and working) of the human clock is still disputed but the hypothalamus region of the brain seems the most likely place.

A comparison of rhythms found on earth with those in other parts of the universe shows links ranging from the obvious to the startling, and it seems likely that this is the area in which the most exciting future discoveries will be made.

Some of these links have been known for thousands of years,

especially those between the positions of the heavenly bodies and the seasons for sowing and reaping. Poets have often written of those affecting the emotions. Chaucer (*c.* 1340–1400) celebrated April as the month 'Than longen folk to goon on pilgrimages', and Tennyson (1809–92) rejoiced that 'In the spring a young man's fancy lightly turns to thoughts of love.'

The precise dating and counting that eventually led to the discovery of more elusive rhythms was little known before the eighteenth century and we can sense in the writings of the early naturalists a feeling of wonder and delight at the unexpected patterns that were coming to light. In 1773, Gilbert White of Selborne told a friend, 'A few house martins begin to appear about the 16 April; usually some few days later than the swallow. For some time after they appear the *hirundines* [swallow family] in general pay no attention to the business of nidification [nest-building] but play and sport . . . About the middle of May, if the weather be fine, the martin begins to think in earnest of providing a mansion for his family . . .'[1]

The Swedish naturalist Linnaeus (1707–78) was so taken with the exactness of the time at which common flowers opened and closed that he devised a 'flower clock' for telling the time of day.

Common sowthistle opens	5 a.m.
Spotted cat's-ear opens	6 a.m.
White water-lily opens	7 a.m.
Scarlet pimpernel opens	8 a.m.
Corn marigold opens	9 a.m.
Chicory opens	10 a.m.
Alpine hawksbeard shuts	11 a.m.
Corn marigold shuts	noon.
Childing pink shuts	1 p.m.
Bearded ice plant shuts	2 p.m.
Pointed hawksbit shuts	3 p.m.
Spotted cat's-ear shuts	4 p.m.
White water-lily shuts	5 p.m.
Evening primrose opens	6 p.m.
Smooth-stemmed poppy shuts	7 p.m.
Day lily shuts	8 p.m.
Night-flowering catchfly opens	9 p.m.[2]

Linnaeus's successors have discovered innumerable such rhythms. Some of the most striking occur in sea creatures whose behaviour is closely linked with the tides and hence with the phases of the moon. If we stroll along a Bermudan beach about an hour after sunset on a summer evening, we may see a luminescence shimmering over the water as thousands of fireworms give off the glow that signals mating. After half an hour, the love light dies and the sea grows dark again. The display takes place once every 29 days during the summer, for the fireworms swarm only at full moon.

Just off Samoa, the moon-triggered mating of the palolo worms brings local fisherfolk an annual feast. The palolos spend most of their lives exploring crannies in the coral reefs but twice a year, the last few segments of their bodies break off and rise to the surface where cells from their reproductive organs fertilize each other. The Samoans hurry out in boats and scoop up the palolo segments literally by the ton. Cooked or raw, they make a highly palatable dish. Although the mating lasts only a matter of hours, the Samoans know exactly when it will take place for it corresponds exactly with the last quarter of the moon in October and November.[3]

There seems to be no good reason why the mating of the palolo should take place at this exact time and, indeed, closely related palolos which live near Japan, off the Malay Archipelago, and elsewhere in the Pacific mate at different stages of the lunar cycle. But they are all predictable, as is another worm called *Convoluta roscoffensis,* a shore-dwelling creature of Brittany and the Channel Islands which feeds on a green seaweed. It has no means of digesting this seaweed but stores it and lives on the substance resulting from the action of light on the seaweed's chlorophyll, itself turning green in the process. When the tide goes out, *convoluta* wriggles to the surface of the sand where it soaks up the light needed for the seaweed to produce these substances. At the turn of the tide, it dives back into the sand for fear of being swept out to sea; hence the dull green patches consisting of masses of tiny *convoluta* worms that come and go with the tides. Indeed, their body clocks are so persistent and so finely adjusted that they work independently of actual tides. If you keep *convoluta* in a tank of damp sand in a laboratory, they will go on boring their way to the surface and burrowing down again at the same times as they would have done were they still living on the seashore.[4]

Simple counting has shown that the sun too influences life on earth in unexpected ways. Far from being a mere ball of fire, to which the planets are tethered by the force of gravity, it has powerful magnetic fields and flings out into space so much energy that it loses more than 4 million tons of mass every second. Visible light rays are only one form of this energy. Others are ultra-violet light, streams of electrically charged particles and radio waves up to a few metres in wavelength.

From time to time, sun spots appear as dark patches on the sun's surface. Really, they are outward-spiralling whirlpools of gas that may cover thousands of millions of square miles. Though cooler than their surroundings, they are still intensely hot and are often associated with violent eruptions of intense brightness that flare up nearby. These flares usually last less than an hour; sun spots for hours, days, weeks, or even months. Together, flares and spots dramatically increase the level of radiation given off by the sun and also the strength of its magnetic fields. Sir John Herschel, the astronomer, discovered as long ago as 1801 that neither spots nor flares occur regularly or at random but in cycles that peak every eleven years.

The full significance of this eleven-year cycle for life on earth is only now being understood. A sunspot 'high' is closely associated with dithering compasses, magnetic storms and brilliant displays of the *aurorae borealis* and *australis*. The flood of ultra-violet rays causes changes in the earth's atmosphere and leads to sudden fade-outs on shortwave radio. When signals do get through, they are much more likely to be marred by 'noise' from direct transmission of radio waves from the sun.

And that is only the beginning. The witch's brew of waves and particles influences weather throughout the world. It has produced an eleven-year cycle of warm wet years which we can trace back more than 600 million years, partly by examining the rings of wood which trees add to their trunks every year and which are wider in warm wet years, and partly by studying varves, the layers of sediment washed into lakes each year from the surrounding hills. These too vary and again are thickest in warm wet years when higher rainfall and increased melt-water from glaciers wash down a higher than average quantity of mud.

This eleven-year cycle affects almost everyone in the world. For sea captains, a 'high' means less icebergs in northern waters, for Hudson's Bay fur trappers an abundance of rabbit skins. A 'low' brings poor vintages in Burgundy vineyards and starvation to millions in the Indian sub-continent. These are, of course tendencies, not certainties, and the cycle itself may vary slightly, but over the ages, the eleven-year cycle of intense solar activity associated with warm wet years on earth has been remarkably constant.[5]

There is also evidence that this eleven-year cycle itself follows a superimposed 'secular' cycle of between eighty and ninety years during which the peaks themselves rise to a maximum before falling back to less extreme heights. This super-cycle has been traced in the frequency of earthquakes in Chile, the number of typhoons in southwest China and even in the first flowering of snowdrops near Frankfurt-am-Main. Lessened solar activity means that westerly winds appear earlier in the year, bringing an earlier spring and therefore earlier snowdrops.

There is evidence too of yet another cycle which the Babylonians called the Saros. It has a period of between eighteen and nineteen years when the moon eclipses the sun at the winter solstice and both bodies exert a pull on the earth in the same direction at the same time. Records of the Nile floods have been kept for some 4,000 years and show definite surges at eighteen- to nineteen-year intervals, as well as at the eleven-year sun spot maxima. Other scientists have suggested a link between earth's recurring ice ages and long-term fluctuations in the sun's heat output. Yet others believe that influences from such planets as Mars, Venus, Mercury, Jupiter, and Saturn may also affect both weather and radio reception.[6]

The human body too responds to this continuous bombardment of energy from space. It follows hundreds of different cycles which regulate our lives in ways which may be trivial or profound. It is not always easy to prove their existence because many of them are obscure and regular observations have to be taken over long periods with elaborate equipment. Often there seems no particular reason for their existence. It has been established, for instance, that we do not breathe evenly through both nostrils at once. For approximately three hours, we take in air mostly through our left nostril and then change to the right for three hours. Why we should do so is a mystery. Even if we do need to change from one nostril to the other, why every three hours? Why not every four, five, or two? Then again, our body temperature is not equal over the whole of our body. One side is always slightly warmer than the other, but there seems no particular reason why this should be so and still less why the left side should be warmer at night, the right during the day.[7]

The most obvious human rhythms are those known as 'circadian', a word that comes from the Latin *circa diem* meaning 'about a day'. It is easy to see that many of our activities, both physical and mental, follow a 24-hour rhythm that is roughly adapted to the alternation of day and night caused by the rising and setting of the sun. Broadly speaking, we tend to sleep during the hours of darkness and stay awake during daylight. Since our patterns of eating, drinking, working, and resting also tend to follow a 24-hour cycle, it seems that most human behaviour follows the sun just as that of *convoluta* and other intertidal creatures follow the moon. It has even been suggested that a regular, daily rhythm of activity is beneficial to human health and that the success of former spas (and presumably too of many present-day health farms) is mainly due not to diet, rest, and 'taking the waters' but to the strictness of the regime which re-establishes natural rhythms which have been put out of joint by the pressures of everyday life.[8]

The systematic study of biological rhythms or chronobiology is comparatively new and it is mostly over the last thirty years that the extent of circadian cycles has become known. They set the pace for innumerable functions, many of them small and little understood but affecting the efficiency of our bodies as a whole. They can mean the difference between life and death. Not surprisingly, there has been an explosion of interest among scientists and laymen alike. According to two leading experts in the field, Dr R. W. T. L. Conroy, Professor of Physiology at the Royal College of Surgeons in Ireland and Dr J. N. Mills, Brackenbury Professor of Physiology at the University of Manchester, England, 'it would indeed be hard to find any physiological measurement which had not in the hands of some assiduous investigator been shown to follow a 24-hour periodicity'.[9]

Though many of these are still disputed, there is wide agreement about others. As long ago as 1845, Dr J. Davy checked his own body temperature every two or three hours from 7 a.m. to 1 a.m. the following morning. He knew that exercise and the temperature of his surroundings would affect his results, so he took care to stay indoors in as constant a temperature as possible and make a point of keeping his level of activity much the same from one hour to the next. He found that his mean daytime temperature was 98·4 degrees Fahrenheit, a yardstick that has proved invaluable as a rough-and-ready guide to 'normality', but his actual temperature fluctuated around this mean by as much as 1·3 degrees, rising to a peak of 98·9 degrees at 4 p.m. and falling to a low of 97·6 degrees at 7 a.m. Since then, many researchers have confirmed his results, with peak temperatures coming in the late afternoon or early evening, and lows around 1 or 4 a.m. It seems that we each have a slightly different pattern from everyone else but it remains remarkably constant even if we stay in bed, run a fever or live in the dark.

Our blood pressure and pulse rate also follow a 24-hour rhythm with peaks in the late afternoons and troughs in the small hours of the morning. Our kidneys too are less active at night, which means that we are not normally awakened by the need to urinate. If we are, the urine passed tends to be acid and low in chloride, potassium, and sodium, compared with our daytime urine which is normally alkaline with peaks of chloride, potassium, and sodium a few hours before or after noon.

The rate at which our bodies burn up carbohydrates and some other substances also shows a daily variation with the lowest points tending to come at night. This is the time when both saliva and gastric juice are scantier and more acid, when our colon goes slow and our brain dawdles. All this happens whether we are awake or asleep. Even if we dance through the night, we shall reach our lowest ebb around 4 a.m. and then gradually perk up so that by 9 a.m. we shall be ready to start a new day.

Men have always known that we reach the zero-point of our vitality towards the end of the small hours. It is the time chosen by secret police to hammer on the door of unsuspecting victims. It is the time when torturers and brain-washers bring their greatest pressure to bear. It is the time when leaders of non-stop encounter groups concentrate hardest on breaking down resistance.

Mostly though, we do not appreciate the extent to which these 24-hour cycles rule our lives. If we were more aware of them and went along with them, we should be able to live both more efficiently and more enjoyably. As it is, we often act flat against them. Most married couples have sex last thing at night, perhaps because they are free of the children and the act of undressing together ensures that the maximum of temptation coincides with the maximum of opportunity. Yet physiologically, this is the least favourable time for sex. Both partners are tired and the man's testosterone level is plumbing the day's low, which comes around midnight. His sexual performance will be at its worst. If they wait until morning, both will be rested and the man's testosterone level will be near its 7 a.m. peak. Sex will be much more enjoyable for both. They are much less likely to have it in the morning, however, partly because the bedroom may be invaded by the children but mainly because it is not customary in our society.

It seems likely too that we may also eat at the wrong times. In recent years, breakfast has tended to become more and more a hurried snack of orange juice, coffee and toast, with the main meal of the day reserved for lunch or dinner in the early afternoon or evening. Yet the evidence suggests that our bodies make much more efficient use of both proteins and carbohydrates eaten soon after waking.

Our bodies also cope more efficiently with alcohol in the early morning but here the implications are less straightforward. At the University of Minnesota, a team led by Dr Franz Halberg took two groups of mice and gave each of them a shot of alcohol, equivalent in human terms to a U.S. quart of vodka. One group took it at night, the other in the morning. Sixty per cent of the morning drinkers died, but only twelve per cent of the night drinkers.[10] It seemed that the mice, which in this respect can be fairly compared with humans, broke down the alcohol more efficiently in the morning and in doing so concentrated its toxic effects. At night, lowered efficiency saved their lives.

An experiment with half a dozen human volunteers told the same story in a different way. They were given small doses of alcohol every hour for two separate periods of thirty-six hours. On both occasions, their blood alcohol levels peaked at 10 a.m., fell steadily until 8 p.m. and then started to rise again. The moral for drinkers is clear. If we take alcohol in the morning or afternoon, it will give us a quick lift but in large quantities is more likely to be harmful. Alcohol taken later at night has milder effects but since it lingers in the body, it is more liable, drink for drink, to get us into trouble with the police. The slower rate of elimination allows higher alcohol levels to build up in breath, blood, and urine.

Knowledge of our body cycles can also improve our efficiency at work, and in a work-oriented world, far more research has been done on work cycles than on any others. The results, though, are not easy to summarise because the term 'work' embraces so many different activities. Humping a sack of flour, doing mental arithmetic, or monitoring a dial, call for widely differing abilities which may peak at different times. Also, students of cycles have tended to categorize work, not people and, as we shall see, people's cycles vary. So in most cases we know only about the 'average' person who may or may not exist.

One of the most comprehensive series of tests ever arranged was that by Dr K. E. Klein of the German Institute for Flight Medicine. The idea was to discover how performance at different tasks varied over the 24 hours and to what extent these could be linked with bodily changes. Heart, blood, breath, temperature, and other tests were taken while volunteers cycled, cudgelled their brains with mental tests, swung violently on tilt-tables or gasped in low-pressure chambers. It came as no surprise that they were at their peak both physically and mentally between 1 and 7 p.m. and at their lowest between 2 and 6 a.m. Reaction times peaked between 2 and 4 p.m., as did skill at tasks requiring co-operation between mind and body. Both sank into a trough between 2 and 4 a.m. This did not necessarily mean that a pilot was always at his least efficient in the small hours. He could, for instance, tolerate a lack of oxygen much better at 3 a.m. than at 3 p.m. Gay Gaer Luce draws the inescapable conclusion: 'As an individual's physiology changes around the clock, he becomes more adept at certain tasks, less adept at others, resistant to some stresses, vulnerable to others. For each challenge, there is an hour at which a person will be best, and an hour when he will be worst.'[11]

Other laboratory tests confirm the German findings and so does research in industry. Teleprinter operators were found to take longest in answering calls between 3 and 4 a.m. and were most prompt in the early afternoon. The records of three meter-readers at a gas works were studied over a period of thirty *years*, during which they worked regular sequences of shifts covering the full period of twenty-four hours both in six- and seven-day weeks. Throughout this time, they consistently made the most number of errors around 3 a.m.[12]

On the whole, field research is probably more reliable than laboratory tests. The people studied know their jobs intimately and do not realise that they are being watched. They work at their own pace, unaware that they are speeding up or slowing down, making more or less errors. There is no element of striving. In the laboratory, tasks are usually unfamiliar to start with and performance tends to improve as subjects learn how to do them. Sometimes, too, they consciously try to overcome the effects of fatigue or biological condition and so distort the results. Even so, research in both laboratory and industry agree in one finding, that performance and body temperature broadly correspond.

Individual temperature cycles vary widely. In 1963, Professor Nathaniel Kleitman, a physiologist at the University of Chicago, published the charts of six men whose temperatures had been taken every two hours throughout the day. Each chart showed a different pattern. Two peaked at 2 p.m. but one of these rose more sharply than it fell while the other fell more sharply than it rose. The first was higher at 10 p.m. than at 10 a.m., the other higher at 10 a.m. than at 10 p.m. Two charts showed plateaux with temperatures tending to remain steady, one from 10 a.m. to 8 p.m., the other from 10 a.m. to 1 a.m. Of the remaining two, one was essentially a plateau lasting from 9 a.m. to 11 p.m. with a slight peak at 3 p.m., the other a plateau lasting from 10 a.m. to 10 p.m. with a slight peak at 2 p.m. and a rather higher peak at 8 p.m.[13]

Each of these men could therefore be expected to have a different pattern of activity and this fits in with everyday experience. We all know people who are night owls, others who are early birds and yet others who fall into a variety of patterns in between. Sir Winston Churchill was a night owl. In his prime, he woke at 9 a.m., spent the rest of the morning reading and writing in bed, rose for lunch and, with the help of an occasional cat-nap, worked until the small hours of the following morning. By contrast, President Harry Truman was an early bird. He was at his desk by 7 a.m. and whenever possible, insisted on an early night.

We can all discover the times at which we work most efficiently by taking our temperatures every two hours during the day for a month or six weeks. As far as possible, tasks that require most effort should be kept for the peaks, routine jobs for the middle ranges and pottering for the lows. Alternatively, we can find the best way of arranging our day by trial and error.

Unfortunately, most of us cannot choose our hours and every factory and office has its night owls who tend to arrive late, blink their way through the morning and get into their stride only in the afternoon. This is precisely when the early birds, who always arrive punctually and drive through the morning at breakneck speed, are beginning to yawn. Both types are a puzzle to the plateau people who go through the day at a steady jog and tend to be irritated alike by the morning breeziness of the one and the apparent laziness of the other.

We have to accept that we can't change our own cycle or the cycles of other people, but by learning to recognize them, we can work more harmoniously together. Getting up late at weekends is a sure sign of a night owl. While he's still snoring, the early birds are waiting outside the supermarket for the doors to open so that they can avoid the mid-morning crowds, who are likely to be plateau types. When going on holiday, night owls drive through the hours of darkness; early birds get a good night's sleep and start betimes in the morning. Plateau types spend their day in bumper-to-bumper traffic jams on busy routes. Night owls have an encyclopaedic knowledge of late television shows; early birds talk knowingly about pre-breakfast broadcasts; plateau types are typically peak-time viewers and listeners.

Employers too can improve the efficiency of their staffs by taking these differences into account. 'Flextime' is perhaps the simplest method. It allows workers to check in and out at times of their own choice provided they put in an agreed number of hours. The early birds will arrive as the night watchman is clocking off; the night owls will stay on until he is clocking on again. Plateau types will tend to work the hours that society regards as 'normal'. All will be giving the best hours of their day to their work. In team projects, it pays to make sure that all members have similar cycles, for otherwise the team as a whole will never function on full power.

Individual cycles should also be recognized on the job. An early-bird boss should not be intolerant of a night-owl employee who spends the morning fiddling with paper clips but becomes so engrossed in his work after lunch that he stays on after everybody else has left. By the same token, employees should work out the boss's type and use their knowledge in timing their approaches for putting up a new project or asking for a rise. Salesmen can also make cycles work for them. They are much more likely to get a positive response from a customer in an 'up' phase if they are explaining a complicated product. If it's a question of wearing down a prospect's resistance, they should pile on the pressure when he is in a trough.

Failure to take account of each other's cycles is a common cause of misunderstanding between husband and wife. The night-owl husband can never understand why his early bird wife is so infuriatingly bright at breakfast but so dreary at midnight when, for him, the party is just beginning. Whenever one of them brings up a family problem for discussion, the other is sure to say, 'You *always* want to talk when I'm feeling tired. You *never* choose the right time.' Unfortunately, the only time right for them both may be around the middle of the day when the husband is away at work. Knowing about cycles won't change them, but it will at least make them mòre tolerant of each other.

It will also help them to cope with their children. Cycles start in the first few months of life and basic types do not change. A child who is lackadaisical in the morning and resists being put to bed at the approved time may well be a night owl whose body clock is out of phase with the pattern of life imposed on him. His parents would be wise not to expect too much of him in the morning and it has even been suggested that bad table manners might be ignored at breakfast and enforced only at the evening meal. One American psychiatrist who recognized the problem in his own child let him stay up for as long as he wanted at night and no harm resulted.

Teachers too should recognize that most children are

biologically incapable of giving of their best equally through the day. Inattention is not necessarily a sign of boredom, laziness or maladjustment. If it occurs regularly at the same time each day in an otherwise hard-working child, his body clock is almost certainly to blame. As a practical measure, it might be worthwhile arranging timetables so that each subject is taught alternately mornings and afternoons. All types will then take at least some of the lessons when they're in an 'up' phase.

Though the pattern of our body cycles stays surprisingly steady over the years, it can easily be disturbed if we cross a number of time zones too quickly. This is a comparatively new problem. In the days when the fastest time from London to San Francisco was some ten days by boat and train, travellers put back their watches an average of approximately an hour a day and their body rhythms readjusted gradually. Now, the journey takes only twelve hours. A businessman leaving London at noon by Greenwich Mean Time (GMT) touches down in San Francisco at midnight GMT when his body clock is urging him to sleep. But San Francisco's clocks are set to Pacific time which is eight hours slow on Greenwich. There, it is four o'clock in the afternoon and his hosts are eager to feed him drinks, take him out to dinner and show him the town. At 2 a.m. Pacific time, *they* drive home to bed and promptly fall asleep. For the visitor, it is now 10 a.m. and he is biologically roused for a day's work. He gets little if any sleep, rises a few hours later and, unless he is a late afternoon type, is already past the day's best. By lunch, his body is ready for dinner; by dinner, it is well past his GMT bedtime. He has to eat, sleep, work and make important decisions at exactly the wrong times.

Research by several drug companies, as well as by the German Institute for Flight Medicine, and the American Federal Aviation Authority, has confirmed that judgment and reaction times both suffer while the body clock is adapting to local time. A transatlantic flight from Oklahoma City to Rome involves a seven-hour time shift and volunteers on an experimental flight were badly affected psychologically for the first day after landing. Mentally, they recovered speedily but it was some four days before their body temperatures and pulse rates were back to normal. Older businessmen took noticeably longer to readjust than young medical students. After a north-south flight from Washington to Santiago, volunteers felt tired for a day or so and reacted more slowly in tests but there was no change in their temperatures or pulse rates. This confirmed that physiological changes are brought about not by the length of time spent on the journey but by the number of time zones crossed.

There is evidence too that the effect of repeatedly crossing time zones may be cumulative. Air-crew constantly flying backwards and forwards across the world may develop eating, sleeping, and digestive problems. Stewardesses may have menstruation difficulties. Sometimes, the symptoms disappear after a holiday but in bad cases, the victim may have to be grounded. Research on mice by Drs Walter Nelson and Franz Halberg of the University of Minnesota suggested that some of the long-term effects might be serious. They took ninety, year-old mice and changed their day-night rhythm every seven days by artificial lighting. These mice lived on average only 88·6 weeks, compared with an average 94·5 weeks of a control group. Upsetting their body-clocks apparently shortened their lives.[14]

Once again, awareness of the problem enables us to alleviate some of the effects. Some airlines now insist that pilots take a rest of at least twelve hours when they gain or lose four hours or more, or when a flight has lasted through 0300 hours local time. Business travellers can often help themselves. Important decisions should be postponed until at least the second day after arrival and some firms in both Britain and America now have rules forbidding executives to sign contracts or make other arrangements during the first twenty-four hours. Ideally, the traveller should spend the week *before* his flight eating, sleeping, and working by local time at his destination. Another method is not to change one's cycle on landing but to continue eating and sleeping at the usual 'home' time, ignoring local time completely.

A more practical solution for the businessman is to arrange a flight arriving at his 'home' bedtime, politely decline either to discuss business or to celebrate on landing, take a sleeping pill and go straight to bed. Anyone who is severely affected by jet-lag may find it helpful to take sleeping-pills regularly (assuming his doctor thinks it wise) until he settles into the new routine and perhaps, too, stimulants at 'low' periods in the middle of the working day. He will be able to do this much more effectively if he has worked out his body cycle beforehand either by taking his temperature regularly or by noting his daily ups and downs.

Shift work is another common reason for disturbed body rhythms. The need to keep ever more costly machinery working for as many hours of the day as possible has brought a growing trend to round-the-clock working in most advanced countries. Though a good deal of research has been undertaken about shift-workers' health and contentment, we know comparatively little about the effect on their body cycles because so many other factors are involved. If a night-shift worker sleeps badly during the day, it may be because his body-clock has not adjusted to the new cycle. Alternatively, his bedroom curtains may not be thick enough to keep out sunlight, the motor-cycle buff next door may spend the morning tuning up his machine or his own children may be home from school and he is tempted to spend the day playing with them.

The effect on health is also arguable. According to Gay Gaer Luce in her exhaustive study of body cycles, 'An unusually high incidence of ulcers has been found among shift workers. Air-traffic controllers usually rotate in shifts, as often as every few days in some airports or at intervals of two weeks in others. Men who cannot endure the schedule are screened out, but ulcers and hypertension are not too uncommon among those who remain. They also have the highest divorce rates in the country.'[15]

Research in Britain also suggests that shift work may be a cause of marital disharmony. After interviews with the wives of 439 shift-workers, the Prices and Incomes Board reported that more than a quarter felt that their husbands' hours were harming their marriages. They complained of loneliness, disrupted sex lives and interference with 'normal leisure and recreational time'. The report concluded that the 'cumulative effect' was 'some reduction in marital happiness and an even greater reduction in the ability to co-ordinate family activities'.[16]

On the other hand, fears that shift-work may affect health are officially discounted in Britain. After a 12-year survey of 8,000 manual workers in ten different companies, the TUC Centenary Institute of Occupational Health found that shift-workers had better attendance records than day workers and that their death rates were almost exactly equal to the national average. The research was supervised by Dr Peter Taylor, former Deputy Director of the Institute. He said: 'Our survey will kill the myth that shift work is bad for you. Possibly about ten per cent of those workers who go on shifts can't adapt themselves, and managements must find some way to help those workers opt out of it. But for the rest there will be no harm.'[17]

Circadian rhythms also have important implications for medicine, enabling doctors to predict the ups and downs of illness, the times at which it can be treated most effectively and even when the major events of life tend to occur. Massive surveys in England, America, and elsewhere have all confirmed what long-suffering G.P.s and midwives had long guessed—that more babies are born at night than during the day. We know now that 10 per cent more are born at night and that the baby cycle rises to a peak between 3 and 4 a.m., plunging to a trough between 1 and 8 p.m. Death too tends to come in the small hours, as we might expect from the downturn in so many bodily functions at that time.

The course of some illnesses is now known to be affected by the fluctuations of our body cycles. Lung capacity falls to a minimum around 6 a.m., rises during the morning and falls again at night, an important consideration for chronic bronchitics. Strokes, the spitting of blood, and cardiac asthma all tend to occur at night. It has long been known that epileptics tend to have their fits in varying patterns but we now know that, taken as a whole, they are more likely to have them in the early morning, especially between 6 and 7 a.m., with a low point between 5 and 9 p.m. and another sharp rise between 10 p.m. and midnight. Those prone to allergies are more likely to suffer late at night. Among the mentally ill, psychotics, depressives, and schizophrenics often show marked changes in the daily rhythm of their body chemical levels. It seemes likely too that most people's vulnerability to bacteria and viruses changes by the hour.

We know that our bodies break down various substances more efficiently at some times of the day than at others. So it is not surprising that the effectiveness of medicines varies with the time at which we take them. If we dose ourselves with aspirin at 7 a.m., the last traces will not leave our bodies until 5 a.m. the next

morning, a period of twenty-two hours. A similar dose taken at
7 p.m. will have disappeared completely after only seventeen
hours. We know too that digitalis has a much more powerful
effect on heart patients when taken at night and in severe cases of
diabetes mellitus, insulin should not be given in equal doses. They
should be varied to restore as far as possible the rhythmic changes
in the 'normal' level.

What we know, however, is almost negligible when we compare it with what we don't know. Experiments on mice and other
animals suggest that anaesthetics, sleeping pills, tranquillisers,
stimulants, and many other drugs also vary in potency with the
circadian rhythms of the patient, and so do vaccines. Our sensitivity to pain and our ability to stand up to the shock of major
surgery soar and plunge through the twenty-four hours. The time
of day at which we are given treatment determines both its
effectiveness and our chances of survival. A sleeping pill whose
effectiveness may last only four hours or so when taken at 11 p.m.
may have a longer-lasting effect when taken at 4 a.m., making us
dangerously accident-prone when we drive to work. At the
moment, our doctors do not have the information needed to
advise us.

So far, we have been discussing circadian rhythms only. These
are the most obvious and apart from women's monthly cycles (see
chapter 4), are those of which we are most conscious. Yet without
thinking, we follow many others from birth to death.

One of these is the ultradian rhythm, i.e. rhythm lasting less
than a day, which first came to notice through Dr Nathaniel
Kleitman's well-known research on sleep. Every ninety minutes
or so, sleeping volunteers wired to recording apparatus in his
laboratory tended to go through a period of rapid eye movements
(REM periods) and if they were roused at these times, they nearly
always told of dreams to which the eye movements related. The
men also reported erections, whether or not their dreams were
openly sexual, and Kleitman suggested that REM periods might
be the peaks of a 'basic rest and activity cycle' involving other
basic drives and continuing through waking hours too. Psychia-
trists Charles Fisher and Stanley Friedman of Mount Sinai
Hospital, New York asked a number of volunteers to spend
between six and nine hours in a comfortable room well supplied

with food, drinks, and cigarettes and watched them through a hidden window. Every ninety minutes or so, each volunteer tended to smoke, drink, take a snack or perhaps even suck his fingers. Other workers have also noted this 90-minute oral cycle. There is evidence too that most of us have stomach contractions at similar intervals, even when we do not eat. We tend to have episodes of day-dreaming approximately every ninety minutes and at these times a rotating spiral, when stopped, seems to go on spinning longer than it normally does. We are also more susceptible to other well-known illusions. Much of this research has been done by Dr Daniel Kripke, director of the Sleep Diagnostic Laboratory at the San Diego Veterans' Hospital, and his colleague Dr Peretz Lavie and they have pointed out the possible advantages of recognising these ultradian rhythms:

> 'Men and women . . . may be able to identify periods of greatest oral drive and plan their meals accordingly. They may compensate for brief periods of physical and mental fatigue by cat napping, or doing something which demands little mental effort. (This ability could be particularly important for people who have to do monotonous jobs requiring a high degree of concentration.)
>
> Couples may even be able to pinpoint periods of enhanced sexuality and use them to make their sexual relationships more harmonious. (On the other hand, "I'm in an ultradian trough" could replace "I've got a headache" as an excuse for avoiding lovemaking.)'[18]

As yet we do not know very much about ultradian rhythms. We know still less about infradian rhythms, i.e. cycles lasting longer than twenty-four hours. Even so, it seems pretty clear that most of us need a short break from our work every seven days, a rather longer one every three months and an extended holiday once a year. Long before Christianity, the peoples of the northern hemisphere celebrated the winter solstice and though the celebration had a religious significance it was also a way of breaking up the cold, grey, dreary winter which in many countries stretches effectively from October to March. Hence the festivities at Christmas and the New Year. Many of us find that we are at our lowest ebb in the first three months of the year but towards the end of March, the sap rises and we get a corresponding peak in births in the following December. Any doctor will tell you that he is much less busy in summer, mainly because we seem better able to resist many of the milder cold and influenza viruses which are ever present but always more potent during our low period in winter.

Many illnesses too have rhythms of their own. Malaria is the most familiar, attacking every second or third day depending on its type, but the new interest in cycles has turned up many others. Psychoses vary enormously but some have been found to be periodic with attacks occurring regularly every two days, once a month or even once a year. Other illnesses which are now known to occur at fixed intervals include a form of peritonitis, oedema (an accumulation of fluid in the ankles and elsewhere), purpura (a blotching of the skin), mouth ulcers, migraines, fever, and hypertension. They can almost always be linked with changes in body chemistry and because of their regularity, can be accurately predicted.

Few doctors, however, are yet alert to the fact that so many illnesses may be periodic. Even if they were, it could be argued that little would be gained. In most cases it would not be economic to spend time working out the periodicity. The diagnosis would be the same and so would the treatment. On the other hand, patients would gain substantially. If we could jot down in our diaries the days on which we could expect to have tummyaches or migraines, we could plan our lives accordingly.

To make maximum use of present-day knowledge we might well need diaries that covered at least eleven years. Well before the Second World War, Maki Takata, a biologist of Toho University, Japan, became famous for his method of calculating the proportion of albumen in blood serum by a process known as the *Takata reaction*. In January 1938, biologists were puzzled by unexplained rises in albumen levels all over the world and every day for four months, Takata tested a patient in Tokyo while a colleague made similar tests on a patient in Kobe. The daily fluctuations in albumen corresponded exactly. Clearly, some factor was affecting patients equally in widely separated places.

Takata spent twenty years checking fluctuations in albumen levels and searching for extra-terrestrial phenomena with a similar pattern of change. At last, he discovered them in the sun. Albumen peaks always occurred when a group of sunspots clustered on the sun's central meridian, so bombarding the earth with the maximum intensity of rays. To confirm his belief that solar radiations were responsible, he flew a volunteer to a height of 30,000 feet where solar radiations have not been filtered through the earth's atmosphere and are therefore much stronger than on the earth's surface. The volunteer's albumen level shot up. As a final proof of the link between albumen level and the sun's rays, Takata showed that albumen levels dropped sharply in people shielded from solar radiation by the moon during an eclipse of the sun. He tested volunteers in 1941, 1943, and 1948, in places where the solar eclipse was total, and got the expected results. It is worth noting that *Takata reaction* tests suggest that ordinary buildings do not shield us against solar radiations that affect albumen levels and the only place where we can normally avoid them is apparently deep under the earth. Tests down a 600-foot mine shaft showed a sharp fall in albumen levels.[19]

If solar radiation can affect the level of albumen in the blood, it seems reasonable to suppose that it may affect the level of other components too. In February 1956, tests in the Soviet Union showed that the proportion of the population with a deficiency of leucocytes or white blood cells doubled from 14 to 29 per cent at a time when there were major eruptions in the sun. In the following year when solar activity passed its maximum intensity, the proportion of leucocytes in the blood of healthy people went up sharply.[20]

The obvious question was: did these changes in the blood affect health? Observations in several countries suggested that they did. In the town of Sochi in the Caucasus, the average number of coronaries is two a day. The day following gigantic solar flares on 17 May 1959, it jumped to twenty. In Pavia, Italy, research over four years showed that the number of coronary attacks closely followed the number and violence of sunspots with some 450 occurring in the peak year of 1958, compared with only 200 in the low year of 1954. Research in France also showed that the number of coronaries corresponded closely with the violence of solar activity. It seems, then, that coronaries are more likely to occur on days immediately following violent solar outbursts and since these reach a peak every eleven years or so, we can expect the number of coronaries to peak every eleven years as well.[21]

Research in Germany, Denmark, Switzerland, and America suggests that cyclic changes in solar activity also affect the number of deaths from pulmonary tuberculosis, attacks of eclampsia in pregnancy, accidents on the roads and in coal mines, the incidence of suicide and the number of patients admitted to mental hospitals.

We now know a good deal about 24-hour circadian rhythms, something about annual rhythms and a good deal less about eleven-year sunspot cycle rhythms. It seems likely that forces from other planets may also affect us, if only in a slight degree at intervals corresponding to changes in their own positions relative to earth and to each other. Of these, we are as yet almost totally ignorant. We can be sure, however, that number will be the key to finding them.

3

Are our Lives ruled by Numbers?

BIORYTHMICS is a system for charting the ups and downs of three separate body cycles which are supposed to control our physical, sensitivity (or emotional), and intellectual states. As Professor Kichinosuke Tatai of Tokyo, Japan said in 1971:

'Unless you spend your days in a vacuum, the tempo of your existence varies a great deal. You may wake with the feeling that this is going to be your day, or experience days when everything you do seems to go wrong. In between these extremes your days will run smoothly and pleasantly.

'If your observations confirm these subtle changes in the tides of life, then you are aware of your Biorhythms, the regulated pulsations in your life system. The convincing part about the Biorhythm theory is that it can be verified by checking our own ups and downs, or through reviews of human error, accident records and death reports.'[1]

Supporters of the theory claim that biorhythmics enables us to work out our physical, sensitivity, and intellectual states (PSI) not just for the present but for any time in the future, so that we need never make the mistake of planning a romantic weekend for a time when we are doomed to feel as passionate as cold potatoes or a mountaineering holiday when our whole body is telling us that we need a rest. Many individuals claim that biorhythmics helps them lead more ordered lives and a number of large commercial firms in several different countries believe that it helps them improve efficiency and cut down accidents. Scientists are divided. Most regard it, at best, as grossly over-simplified but some, who are equally well qualified, give it their ardent support.

It springs from a monograph published in 1897 with a title all too reminiscent of an undergraduate spoof: *The relations between the nose and the female sex organs from the biological aspect.* Its author, Dr Wilhelm Fliess, was as serious as every other enthusiast who believes that he has found the key to the universe. He was a Berlin nose-and-throat specialist who had a passionate friendship with Sigmund Freud during the 1890s and though far from being Freud's equal intellectually, he had an imaginative flair which his friend found irresistible. He was deeply interested in the new psychology and agreed with Freud that there was a feminine element in every man and a masculine element in every woman. He held too that one kind of neurosis had particular connections with the nose, a view based largely on the fact that it swelled during sex and menstruation. The symptoms of this 'nasal reflex neurosis', which was similar to what Freud called an 'actual neurosis', included headache, backache, and digestive upsets. All of them, said Fliess, could be relieved via the nose by a sniff of cocaine which was then in vogue as a sedative.

As if that weren't enough, Fliess used the established link between the nose and menstruation for a flight of fancy that carried him all the way to the stars. If a woman's nose followed the 28-day menstrual cycle, he reasoned, so too must a man's, for every man had a feminine element. Moreover, since the nasal reflex neurosis proved that there was a link between the genitals and many other organs, these too must be affected by the 28-day cycle in both sexes. With the bit firmly between his teeth, he soon ran down a second cycle of 23 days corresponding, for no particular reason, to the interval between the end of one menstrual period and the beginning of the next. This must, insisted Fliess, be a male cycle but like its 28-day counterpart operated in both sexes from birth and went on even beyond the grave. He wrote:

> 'Recognition of these things led to the further insight that the development of our organism takes place by fits and starts in these sexual periods, and that the day of our death is determined by them as much as is the day of our birth. The disturbances of illness are subject to the same periodic laws as these periodic phenomena themselves.
>
> 'A mother transmits her periods to her child and determines its sex by the period which is first transmitted. The periods then continue in the child and are repeated with the same

rhythm from generation to generation. They can no more be created anew than can energy, and their rhythm survives as long as organised beings reproduce themselves sexually. These rhythms are not restricted to mankind, but extend into the animal kingdom and probably throughout the organic world. The wonderful accuracy with which the period of 23, or, as the case may be, 28 whole days is observed permits one to suspect a deeper connection between astronomical relations and the creation of organisms.'[2]

Fliess's theory now strikes one as slightly dotty. Apart from the link between nasal congestion and menstruation, almost everything was conjecture and though, as we shall see, some of his surmises were not so wide of the mark, they were totally unjustified in the light of existing scientific knowledge. His notion that sex is transmitted with periods is, of course, rubbish.

Freud, however, was impressed. He was worried about his father and asked Fliess to work out from their respective birth dates whether Freud senior would survive Bismarck. He 'could have shouted with joy' when Fliess told him that he was on the brink of discovering a contraceptive safe period and wondered whether two 'actual' neuroses discovered by himself might not be precipitated by bodily secretions released at intervals of 23 and 28 days. He dubbed Fliess 'the Kepler of biology' for his attempts to link the heavenly bodies with earthly life and wrote:

'Yesterday, the glad news reached me that the enigmas of the world and of life were beginning to yield an answer, news of a successful result of thought such as no dream could excel. Whether the path to the final goal, to which your decision to use mathematics points, will prove to be short or long, I feel sure it is open to you.'[3]

It was not until 1899 that Freud began to have doubts. For one thing, Fliess was now following the path of nearly every numerologist in history. When he came across facts that did not fit his theory, he made it more and more complicated to accommodate the new facts. He busied himself with multiples of 23 and 28 days. He added them together and found a third period of 51 days which could itself be multiplied. He discovered too a $7\frac{1}{2}$-year cycle which governed men's fortunes.

As the friendship cooled off, Freud realized that a psychology based on numbers could never be reconciled with his own view that human behaviour was profoundly affected by the dynamic of forces hidden in the unconscious mind. Fliess insisted that psychological changes could never be attributed to the effects of psychoanalysis alone. After a last meeting at Achensee, Austria in the summer of 1900, the friends parted for good and though Freud admitted as late as 1924 that there was an unknown factor in the workings of the pleasure-unpleasure principle, 'perhaps something rhythmic, the periodical ringing of the changes, the risings and fallings of the stimuli', he was plainly scornful of Fliess's ever more convulsed numberings. In 1906, when his friend and colleague Ernest Jones asked him how Fliess explained the interval between two attacks of appendicitis when it was a multiple of neither 23 nor 28, he replied: 'That wouldn't have bothered Fliess. He was an expert mathematician and by multiplying 23 and 28 by the difference between them and adding or subtracting the results, he would always arrive at the number he wanted.'[4]

Fliess was undeterred by Freud's desertion. His book *Der Ablauf des Lebens* (*The Course of Life*) brought him a brief notoriety in 1906 and Hermann Swoboda, a Viennese philosopher, claimed that his 28-day period could also be found in dreams. Though most scientists rejected his work out of hand, his number system continued to appeal to a minority, and enthusiasts in several countries attempted to prove it experimentally. In 1927, Dr Rexford B. Hersey of the University of Pennsylvania studied a group of 25 railway repair workers who had more than their fair share of accidents. After plotting not just their mishaps but also their moods for a whole year, he claimed to have found that each had a 'low' every $11\frac{1}{2}$ days and at these times was more accident prone. The $11\frac{1}{2}$-day cycle was exactly half Fliess's 23-day cycle and seemed to show that accident proneness was linked with the male physical cycle.

Research by Dr J. Bennet of Doctor's Hospital in Philadelphia seemed to confirm Hersey's findings. After checking the records of thousands of patients, Bennet too found a physical cycle of 23 days with two lows, one at the beginning of the period and another halfway through. Meanwhile, in the early twenties, Dr Alfred Teltscher of Innsbruck, Austria had studied the performance of 5,000 school children and claimed to have discovered a third cycle of 33 days which regulated intellectual activity. After further work in Switzerland, the system became known as biorhythmics.

The early history of biorhythmics was distinctly sketchy for interest in it was, appropriately, cyclical, though no one has yet worked out the length of the cycles! Since the early 1960s, however, it has attracted world-wide attention, partly because of increased scientific interest in biological time cycles and partly through the efforts of Professor Kichinosuke Tatai of Tokyo, Japan. Let there be no doubt about Professor Tatai's standing. He is a graduate of Tokyo Medical School (1940) and of Harvard School of Public Health (1952). He spent twenty years lecturing on human factors in public health at the Institute of Public Health, Tokyo and is currently professor of human physiology in nutrition at Tokyo University. He lists his degrees as M.D., M.P.H., and Dr Med. Sci. and apart from memberships of numerous Japanese and international medical societies, is also Japanese representative on the International Association for Suicide Prevention.

Tatai came across biorhythmics while studying human body cycles in the 1950s and like most scientists was highly sceptical. He felt it was 'too mathematical to be physiology'. Even so, the peculiar appeal of number systems seems to have stirred his imagination and for two years he kept 'very careful records'. His final analysis suggested that biorhythmics was accurate for at least 70 per cent of the time, a proportion that 'was certainly meaningful when compared with the scientifically based weather forecasts which used only to be 58 per cent correct in Japan.' After further work, Tatai formed the Japan Biorhythm Laboratory, of which he is now director, wrote a popular book, *The Mysterious Rhythms of Life* (available only in Japanese), and lectured widely on the subject. His motive was, he said, 'to enhance what I like to call the dignity of life which is in my opinion, as a doctor of medicine, the basis of human welfare. Unfortunately, in the world today there is a tendency to disregard this dignity but I am of the opinion that if everyone were to keep this natural law of human nature in mind and to control their behaviour by this training, most human tragedy would disappear.'[5]

The system, as refined by Tatai is called Biorhythm—PSI, the initials standing for the three functions—physical, sensitivity, and intellectual—ruled by the 23, 28, and 33 day cycles on which it is based. They start at the moment of birth and continue regularly throughout life. Each is divided into two equal phases which can be represented by a graph (Figure 1), the upper part consisting of positive days on which the function is working strongly and the lower, regenerative phase when it is recuperating its energies. The most important days to note are those when we are crossing from one phase of a cycle to the other. They are called 'critical' because they are those on which we are most liable to have accidents, feel irritable or make mistakes. Though they peak for a period of twenty-four hours, there is no sudden cut-off and their influence spreads for two days on either side (see Figure 2). Nor does the 24-hour period itself necessarily correspond to a single calendar day. In the odd-numbered cycles it is usually spread over two and since the cycles start at the moment of birth, a baby born at 11.55 p.m. on 1 January will have criticals co-inciding almost exactly with those of a baby born at 12.05 a.m. on 2 January. Another point to notice is that our cycles, being of different lengths, are often in different phases. Our physical may be positive while both sensitivity and intellectual are regenerative. We rarely get two or three criticals coinciding but when we do, we are highly vulnerable.

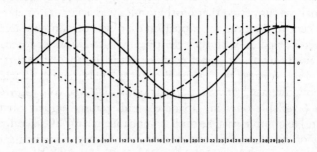

FIGURE 1

———————————— Physical curve

– – – – – – – – – Sensitivity curve

. . . . Intellectual curve

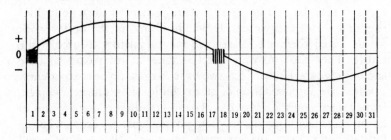

FIGURE 2

Thirty-three day intellectual cycle

The theory is simple but it is not so easy to apply. First comes the practical problem of working out the present and future state of cycles that started twenty, forty, even seventy years ago. We can, of course, get out pencil and paper, calculate the number of days we have been alive (not forgetting the extra days for leap years) and divide the result by 23, 28, and 33 to find the current position, but even on an intellectually positive day, few of us would get it right. It is not surprising, therefore, that a minor industry has grown up in pocket calculators that do most of the work for us. They come in the form of curves, slide-rules, and rotating discs, most of which cost several pounds and have to be used in conjunction with elaborate tables. In Britain, the Bio-rhythmic Research Association sells individual computer print-outs covering a whole year and the Swiss have designed a watch which can be adjusted to the wearer's birth date to show his biorhythmic state day by day. It costs £89.95 (at 1976 prices), which leads one to wonder whether it is a mere coincidence that Switzerland, which is so heavily dependent on its clock and watch industry, is also one of the leading countries in biorhythmic research.

The second snag about biorhythmics is the nature of the information it claims to give us. Not even its keenest advocates suggest that it works for everyone all the time. Norman Chipping, a practising dentist who is also director of the Biorhythmic Research Association, admits that one person in five has one or more cycles which lag behind or march ahead of the standard pattern. He proposes that before looking into the future, we keep a diary of our physical, sensitivity, and intellectual ups and downs for six months. Only after checking them against our biorhythmic states can we decide which, if any, of the cycles work for us. Even then, we cannot use our calculators to predict the future. They show us only our PSI states at any particular time and to judge the likely effects, we have to draw on past experience.

Often, the interpretation of biorhythmic states is more complex. It may seem contradictory that Tokyo taxi-drivers were found to be more accident-prone on their physical-critical days, whereas long-distance lorry drivers were most at risk on intellectual criticals. Students of biorhythmics explain that driving a taxi in a busy city is physically tiring and demands quick reactions. It follows that a driver who is physically below par will tire sooner, react less quickly and be more liable to accidents. Long-distance lorry driving, however, is largely a matter of watch-keeping, especially on motorways. Quick reactions are rarely needed and manoeuvres can usually be planned at leisure. Judgment is the quality most called for and drivers are most affected by their intellectual cycle. How ordinary motorists should interpret their biorhythmic states depends very much on their individual reactions and the type of driving they do. Assuming that they are influenced by all three cycles, it would make sense to drive more slowly on physical criticals, avoid complicated manoeuvres on intellectual criticals and keep out of arguments with other drivers on sensitivity criticals. When two or three criticals coincide ('double or treble criticals'), it would be safer to leave the car in the garage!

Biorhythmics, it is claimed, can also help us use our talents more effectively. An ergonomics student found that he was best able to absorb new knowledge when he was in an intellectual 'plus' phase, and a physical 'plus' also helped, but when his sensitivity cycle was also positive, his mind was so alert that he was unable to concentrate. He could learn new facts more easily in a sensitivity minus phase. He found too that he made *less* errors on intellectual criticals because he knew they were likely to occur and double-checked his answers.

Similarly, families can avoid rows by making allowances for each other's states, not just criticals but some others too. Extroverts, for instance, are said to be much easier to live with when they are at the bottom of their sensitivity cycle and therefore less bouncy. We can also make allowances for ourselves. If we are accidentally jostled on a day we know to be a sensitivity critical, we can curb our irritation and keep the incident in perspective. During a physical minus phase, we need not feel despondent if we tire sooner than expected on a long walk. Anyone who believes in biorhythmics can, at least in theory, take a more philosophical view of life generally.

The question is, does it work? Or is it one of those mildly paranoid delusional systems which numerologists of all kinds are liable to conjure out of the air?

Its supporters claim that many accidents, tragedies, and disasters have occurred on days that were biologically unfavourable for the people involved. According to Professor Tatai, 241 people were killed in three air crashes that took place in Japan during the month of July 1971. One was a Towa Domestic Airways YS II whose pilot had an intellectual critical at the time. The others were a P2 V7 whose pilot was critical both physically and intellectually, and a Boeing 707 hit in mid-air by a fighter piloted by a student under instruction. The student was in an 'almost' triply critical state, the instructor in an intellectually critical. Other instances are quoted from sport.[6]

Many of these claims are not easy to check but when we look at the deaths of some well-known personalities we are on firmer ground. The American actress Diana Barrymore was born on 3 March 1921 and committed suicide on 25 January 1960. She was within two days of criticals in both her intellectual and sensitivity cycles. When Clark Gable (born 1 February 1901) was in hospital after a heart attack in November 1960, George Thommen, an American supporter of biorhythmics warned on the Long John Radio Program, 'If I were Mr Gable's doctor, I would have everything that I possibly could have in his room for the greatest emergency on the 16th of this month.' Clark Gable died on 16 November 1960, just as Thommen had hinted. It was a physical critical for Gable and also within two days of a sensitivity critical.

These cases seem to confirm biorhythmic theory, as anyone can check for himself. I wondered, however, whether they were exceptional and made two random checks of my own. The first was on Kenneth Allsop, the British author, journalist, and television interviewer who was born on 29 January 1920 and took an overdose of sleeping pills on 23 May 1973. Using Professor Tatai's tables, I quickly found that he was within one day of his sensitivity critical when he died and was coming down to the lowest point of his physical cycle.

My second check was on the worst air disaster in British history: the crash of the British European Airways Trident Papa-India shortly after take-off from London's Heathrow Airport on 18 June 1972 with the death of 118 passengers and crew. The official enquiry showed that the captain, Stanley Key, had severe atherosclerosis or 'hardening' of all three coronary arteries and that bleeding from a tear in the lining of one of these had started between two hours and one minute before he died. Captain Key had been emotionally upset immediately before the flight and there was evidence of a row in the crew room.

The report went on: 'Since this event took place about an hour and a half before takeoff, it is not unreasonable to suppose that it may have been the cause of the initial rise in blood pressure which resulted in the rupture of the small blood vessel. The process, thereafter, would be gradual but dynamic, culminating in the separation of the lining of the arterial wall at a time which medically at least can only be a subject of surmise but may have been in the last thirty seconds of life.' The result was a series of flying errors which made the crash inevitable.[7]

After ten minutes' research in the public search department of the Registrar-General's Office, I discovered that Captain Key was born on 28 March 1921. The day of the crash was therefore a physical critical and the following day an intellectual critical— exactly the state which, according to biorhythmic theory, would be associated with his bodily vulnerability and consequently impaired judgment. If B.E.A. had taken the view that no one in his biorhythmic condition should be allowed to carry out duties in which he might endanger his own life or the lives of others, the crash would never have happened. (In fairness to B.E.A., however it must be made clear that biorhythmics is not an accepted science and they would face serious criticism, perhaps even ridicule, if they grounded all aircrew on unfavourable days. Moreover, Captain Key had passed routine medical checks and no one had any way of knowing about his long-standing heart condition.)

Such examples are striking but they are not conclusive. Everyone has half a dozen criticals a month and if we took all the heart attacks, suicides, accidents, sporting reversals, and emotional outbursts occurring on any particular day, we should expect to find on chance alone that it was a critical for about one person in five. To make out even a *prima facie* case for biorhythmics, we should need to show that a significantly higher proportion of the victims were in a critical state, an impossible task unless we limit the investigation to a particular category of incidents in a particular place at a particular time.

Traffic accidents are one possiblity. We must, of course, take only those for which a single driver was responsible and exclude those caused by mechanical defects, abnormal weather conditions, and other outside circumstances. In 1954, Dr Reinhold Bochow of Humboldt University, Berlin investigated 497 'driver only' accidents and found that 'the majority' happened on criticals or double-criticals. Two years later, Chief Engineer Otto Tone of Hanover investigated traffic accidents in the city and found that 83 per cent could be linked with criticals.[8]

Perhaps the most thorough work in recent years, however, is that undertaken by Professor H. L. le Roi of the biometry laboratory of the Technical University of Zurich, Switzerland, in 1970. The Swiss Accident Prevention Officers' Conference gave him detailed information on 2,468 accidents, together with the birth dates of the drivers, who were all considered 'totally responsible'. For various reasons, nine were eliminated, leaving 2,459. Their charts were worked out in two cycles only, physical and sensitivity, giving criticals on days 1, 12, 13, and 23 and on days 1, 14, 15, and 28 respectively. The proportion of accidents that could theoretically be expected on these days was 29·19 per cent. The proportion that actually happened was 30·87 per cent, which represented in biological terms, 'a significant difference.'

Professor le Roi concluded that drivers tended to have more accidents on their critical days, a conclusion with a probability of error of slightly under 0·04. Previous studies of driver-responsible accidents in the cantons of Zurich, Bern, and Solothurn and in the city of Zurich showed that a slightly higher proportion had

happened on physical or sensitivity criticals—30·99 per cent. This
was rated as 'highly significant.' Taking the two groups of material
together, Professor le Roi concluded: 'It cannot be ignored that
biorhythm, i.e. the biorhythmic condition, bears a definite
relationship to accident frequency.'[9]

As so often happens in biorhythmics, it doesn't sound very
convincing. The influence is so small, far smaller than we should
expect from Professor Tatai's estimate of 70 per cent accuracy or
the B.R.A.'s statement that biorhythmics works for eight people
out of ten. Obviously, we do not expect everyone to have an
accident on every critical or to avoid accidents completely on
other days. Even 'driver only' accidents may be associated with a
hangover, a marital row, the effect of sleeping pills or a dozen
other factors. If the theory is to have any practical merit we might
expect a much closer correlation of accidents with criticals than
Professor le Roi found.

On the other hand, his results may simply mean that a proportion of Swiss drivers are sufficiently aware of the biorhythmic
hypothesis to drive less often and more carefully on their critical
days. This does not necessarily mean, of course, that the hypothesis is valid, only that drivers believe in it or at least think it
worth paying some attention 'just to be on the safe side.' Unfortunately, we do not have enough evidence to be sure either way.
An experiment designed to investigate decisively the claims of
biorhythmics would have to be complicated and subtle, taking
this and many other factors into account.

There is, though, strong evidence that a *knowledge* of biorhythmic states can help to prevent accidents. In Japan especially,
a number of large organizations have worked out the charts of
workers on jobs in which accidents are a real danger. One was the
Yokohama Kita Telegraph and Telephone Office which in April
1967 ordered its motor-cycle messengers to fly a yellow pennant on
their double criticals and a red pennant on triple criticals to remind
them to take special care. Over a period of three years, the
number of accidents was cut to a quarter of the previous level, the
only serious mishap being a collision with the rear end of a car
after 800,000 kilometres in April 1969. Between then and 31 May
1973, messengers covered another 1,879,847 kilometres without a
single serious accident. Among other Japanese companies which

have successfully used biorhythmics to cut traffic accidents are the bus branch of the Omi Railway Company, the taxi section of the Kokusai Automobile Company, Tokyo, the Asahi Mutual Insurance Company, Tokyo, and the Nissan Motor Company of Yokohama. The last two companies claim too that biorhythmics has increased sales.

Mr Russell K. Anderson, former president of R.K. Anderson Associates, an industrial loss prevention consultancy of Rutherford, New Jersey, claimed that an analysis of a thousand accidents over two years had shown that more than 90 per cent occurred on criticals. He told delegates to an American safety convention how they could use biorhythmics to cut accidents in their own factories:

'I cannot say that if you have 5,000 or even 500 employees in your plant that you should prepare charts for all of them. We use biorhythm in all places where there is a possibility of a far-reaching and serious accident. The chemical industry, the nuclear field, or any place where there is danger of a serious catastrophe. We use it in cases where we have executives subject to extreme stress, and particularly in those cases where we have already seen some signs of physical depreciation in the individual.

'The airlines, private pilots, railroad engineers, motormen, truck drivers, and salesmen driving cars all day are ideal subjects for biorhythm studies. . . .'

'Our practice in the use of biorhythm is that we get a complete list of all those people we feel should be checked. We make up the charts for the months ahead. These charts are turned over to the plant supervisor who handles personnel problems in the plant. These charts are studied by the plant supervisor, and he makes the reports for each department which are handed over to the foreman. These reports give the man's name and a statement similar to this: "Employee Jones, low intellectually, high physically, and emotionally critical." A supervisor reading this report for the day will automatically watch this man as the factors here could lead to an accident. The fact that the man is low intellectually means that he might not be aware of a hazard, his thinking will be slower, his reaction time slower than normal. The high physically indicates that he will be able to handle physical labour easier, such as lifting, pulling, or any other feat that requires manual strength. However, being low intellectually and critical emotionally he might well exceed the strength required for the job with the result that he might not use his strength properly and the accident might well result.

'The man will be watched by his supervisor just in case a situation might occur wherein the conditions just described be taking place, and the supervisor seeing it might stop the man before it became serious.

'When a man has a double or triple critical day or period, he must be watched closely to see that he does not do the thing that will lead to an accident.'[10]

A sceptic might say that biorhythmics is a copybook example of a 'heads I win, tails you lose' theory. If I fall off a cliff on a triple critical, the triple critical is to blame. If I don't fall off a cliff, it's because I knew about the triple critical and took special care. Either way, biorhythmics is 'proved' right.

Besides, working out the monthly charts of a fleet of drivers costs time and money and would be undertaken only by a company exceptionally keen on safety. Mr Hiroyasu Innami, Acting Manager of the Group Insurance Department of the Asahi Mutual 'emphasizes that the record card only produces useful results when it is used together with other means of preventing accidents . . . The practical action taken is only to give advice to employees on their critical days but there is no change in duties.'[11] Though other reports do not say so, it seems extremely unlikely that use of biorhythmics is the only extra precaution taken and we should expect, for instance, other measures such as the limitation of working hours, speed restrictions, more careful selection and training of drivers, and improved vehicle maintenance. Even if these were already in hand, the introduction of biorhythmics would bring a new awareness of possible dangers and the general attitude to safety measures would become that much keener.

Constant checking on their biorhythmic states and regular reminders by word, by coloured pennant or perhaps even by being taken off the job at double or triple criticals would make drivers highly safety-conscious even if, objectively, biorhythmics had no effect whatsoever on their performance. It would be a psychological placebo, and none the less effective for being so. Once again, conclusive proof eludes us and we are left with the feeling, 'yes, but . . .'

Despite the lack of generally accepted scientific proof, claims are being made for biorhythmics in many fields. In England, Dr Keith E. Jolles, M.R.C.S., L.R.C.P., a Birmingham general practitioner, consultant in motoring medicine and medical adviser to the B.R.A., kept biorhythmic records of an eleven-year-old girl suffering from grand-mal epilepsy. Although taking phenobarbitone, she had seven fits in nine months and six of them were in positive phases of her intellectual cycle. As we are all in the positive phase of our intellectual cycle for almost half of every 66-day period, the result might be chance but if not, it might mean that epileptics could reduce the number of fits by taking extra drugs at these times. Dr Jolles emphasized the difficulties of experimenting on sick people:

> 'Often the interests of the researcher and the clinicians clash, but there should be no question as to which should take precedence. As the phenobarbitone had not been sufficient to suppress the fits, phenytoin was added to her treatment, since when there have been no further fits for some months. Since that is the object of treatment, no more can be done here. But if in other cases the biogram can again show that attacks are most likely during certain periods, then it would seem reasonable to try and not only suitably modify the patient's general management at those predictable times of risk, but also step up the dose of anti-convulsant drugs for the duration of each danger period only.
>
> 'I can hardly wait to put some cases of migraine through this simple biorhythmic investigation.'[12]

Dr Jolles also wrote:

> 'I believe that medicine and surgery will provide one of the most fruitful fields for the practical application of bio-rhythmic studies. It is known already that a patient's pain threshold is raised during his physical plus phase (dental

operations tend to be less painful during that period) and wounds tend to heal more rapidly, for example after surgery. In elective surgical procedures, especially in the case of major surgery, this could become a valuable factor in deciding upon the best timing.'[12]

In Switzerland, Jack Gunthard, the national gymnastic coach has used biorhythmics in training, choosing teams, and assessing opposition. Helmut Benthaus, trainer of Basle football club, said that he was 'convinced that with Biorhythm, the responsible and conscientious trainer has a means by which he can increase the performance of the players entrusted to him.' In Japan, teachers can buy a device called a Bio-conditioner which enables them to keep a continuous watch on their pupils' cycles so that they can present new information at times when it is most likely to be absorbed, though no one has yet explained how they can do this with a class of thirty pupils whose cycles are all different. There is even a theory that women have a 70 per cent chance of bearing a male child if they conceive when their physical cycle is positive and their sensitivity cycle in the regenerative. When the phases are reversed, there is said to be a similar chance that the baby will be a girl.[13]

How can we sum up the present status of biorhythmics? According to Dr Jolles: 'that awareness of one's critical days can lead to dramatic reductions in accident rates and yield many other benefits has now been established beyond doubt in Japan, Europe, and the U.S.A. (including the U.S. Atomic Energy Commission). We have the authenticated reports.

'Biorhythmics is not a gimmick. It is not, nor claims to be, some kind of astrology, horoscope or other way of foretelling the future. It does not offer instant happiness, success or safety. But it is a perfectly serious scientific study which, properly used, may enhance these qualities of life.'[14]

Since the airline Swissair has often been quoted as an organization that was seriously interested in biorhythmics, I asked their medical director, Dr H. Gartmann for his views. He wrote: 'The only thing we have done is to study carefully and without any prejudice the theory and practice of biorhythms. Our final conclusion: pure superstition and nonsense.'[15]

Dr Andrew Ahlgren, Associate Director, Center for Educational Development, Academica Administration, Minneapolis, attacked biorhythmics on three separate counts. First, the anecdotal evidence. 'The list of relatives' accidents and illnesses is of course pure hokum by which any crackpot theories can be "proved" (and for millennia have been).' Secondly, the method of interpretation. 'The flexibility of explanation gives them almost complete inventive freedom. (If your mental curve is up and your physical curve is down, you might have an accident because you are too "exuberant" for your body; if your physical curve is up and your mental curve is down, you might have an accident because your body "outruns" your judgment; if both are in the middle, then you have a "double zero point" [double critical] and you might have an accident because you are "out of gear" and "vulnerable".)' Thirdly, in a criticism of actuarial research carried out in the 1930s by Dr Hans Schwing of the Swiss Institute of Technology, Dr Ahlgren points out the weakness of a case based on the correlation of accidents with criticals. 'He apparently scanned data on accidents from insurance companies and found that 60 per cent of accidents occurred when victims were at emotional or physical "zero points". How amazing the result is depends of course on his methods and on what "at" means in "at a zero point". (Note that about half of *all* days are within a day to either side of a zero point on one or more of the three curves.)' Summing up, he challenged the very idea of any inflexible scheme of cycles, for all generally accepted body cycles change. 'Variations in rhythms caused by development, disease or even travel would suggest that invariance over a lifetime is very unlikely. Perhaps even more to the point is the gradual emergence of regular rhythms in early childhood, making very untenable a phase fixed at the precise date of birth.'[16]

Dr Ahlgren's arguments are cogent and the overwhelming majority of medical and scientific opinion, at least in the West, would go along with them. We could add too that biorhythmics seems to take no account of circadian and other body cycles whose existence has now been established incontrovertibly. Yet with all these qualifications, it is hard to dismiss out of hand the work of Professor Tatai and the actions of businessmen who have put their money where their numbers are. The anecdotal evidence, though suspect by the criteria of science, is sometimes too striking for a layman to overlook entirely. Could it be only coincidence (whatever that may mean) that Clark Gable should have had his heart attack on a critical or that Captain Key should collapse at the controls of the Trident on a double critical? If we take a rigorous view, we must agree with the objections but the inner voice still gnaws at our logical deduction.

In many ways, the position of biorhythmics today is very much like that of acupuncture twenty years ago. The overwhelming majority of informed Western opinion dismissed the system of meridians and pulses as Oriental quackery. Now, majority opinion is swinging to the view that acupuncture sometimes works, though no one knows why. Biorhythmics *could* prove to be numerological nonsense. On the other hand, further research may prove that it too works for reasons that have yet to be discovered.

4

Why is 13 unlucky?

WHY do so many of us feel that 13 is unlucky?

I say 'feel' rather than 'think', for few of us are as certain that
13 is unlucky, as we are that 12 is 4 times 3 or that there are sixty
minutes in one hour. Indeed, a recent survey showed that only
one person in twenty admitted to the superstition.[1]

Yet our actions belie our words. Triskaidekaphobia, or fear of
13, costs Americans an estimated £500 million a year in absenteeism, cancelled appointments, and loss of business on days of the
month numbered 13. Editors of popular newspapers know that
'unlucky 13' stories will always have an avid readership. When
the Apollo 13 mission was reported in 1970, correspondents were
quick to note that it was launched at 1313 hours American Central
Time from pad 39, which is three times 13. Three of the sleeping
periods arranged for the astronauts were timed to start at thirteen
minutes past the hour and so was one of the possible splashdown
times. The story took an added drama when the first of several
set-backs occurred on April 13.[2]

It scarcely needs saying that there is no evidence whatsoever
that 13 is more or less unlucky than any other number. Yet so
many people *feel* it is that anyone dealing with the public must
take these feelings into account. In many hotels, floor and room
numbers jump from 12 to 14 and so do seat numbers in planes
flown by some airlines. At Geneva international airport in
Switzerland, the hours marked on the 24-hour clock ran '. . . 12,
12a, 14 . . .' and 13 was omitted from the numbering of arrival and
departure gates. In Italy, it is usual to leave out 13 in the numbering
of lottery tickets and in France, hosts who find themselves with
a party of thirteen can hire a professional *quatorzième* from an
agency. At the Johns Hopkins Hospital in Baltimore, Maryland,
there is no operating theatre numbered 13. A hospital spokesman

was quoted as saying: 'Patients have enough anxiety without seeing that number above the door as they are wheeled in.'³

The taboo operates at all social levels. We have all read of low-income families who ask for their street numbers to be changed from 13 to 12a. At the other end of the scale, Paul Getty, the American multi-millionaire, admitted, 'I wouldn't care to be one of 13 at table.' During the royal tour of Germany in 1965, officials at Duisburg railway station noticed that the Queen of England's train was scheduled to leave from platform 13. They changed it to 12a. Thirty-five years before, when her younger sister Princess Margaret was born at Glamis Castle in Scotland, palace officials did not register the birth immediately because the next registration number was 13. They waited three days until another baby had been allotted the number 13, before registering the Princess's birth as number 14.⁴

The reason why 13 makes so many of us feel uneasy has never been satisfactorily explained. Rationalists insist that it is the result of conditioning. *Because* we have been brought up to believe that 13 is unlucky, they say, we take particular note of the occasions when bad luck follows, ignoring those when it does not. We may even become so jittery on Friday 13th that we do in fact have more accidents, not because of the inherent power of 13 but because of our attitude towards it. Professor Gustav Jahoda of the University of Strathclyde in Glasgow, Scotland wrote: 'This kind of process has been dubbed "the self-fulfilling prophecy", and is not uncommon in human affairs. For instance, to take an uncomfortably practical example, if foreign investors believe that the £ is going to fall, they will sell their holdings, which will cause the £ to fall.'⁵

This may be true but it is irrelevant. It fails to explain why we came to regard 13 as unlucky in the first place. Why not 12 or 14? Why not 7 or 359? The self-fulfilling prophecy of the falling £ is based on prior experience: the unreliable performance of the £ in the world's money markets ever since the Second World War. What has 13 ever done to deserve its stigma? Professor Jahoda does not tell us, nor do other behaviourists. Until they do, their explanation is no explanation at all.

Traditionally, the belief that 13 is unlucky has been traced back to Jesus and his twelve disciples who made up 13 when they sat down to the Last Supper, the inference being that at least one of any party of 13 eating together will shortly die. Others quote Scandinavian mythology. The demonic Loki gatecrashed a banquet held by the other gods, making 13 in all. He stirred up trouble, was thrown out and later brought about the death of Balder, god of light. So again one of a company of 13 perished.

It has always seemed to me that these two stories do not come anywhere near to explaining the fear of 13. Most people have never heard of Loki and few of us associate the Last Supper with 13. In neither case was the number significant in causing the tragedy, nor did it have any symbolic meaning. It was merely a coincidence that went unnoticed at the time and never carried an emotional charge powerful enough to cause the gut feeling about 13 which centuries of reason, and rationalization, have been unable to overcome. It seems much more likely that both 'explanations' are further projections of our anxiety.

So what *does* cause that anxiety? I believe that there is no simple answer. Like every number, 13 is a symbol. If we look into mythology, religion, history, and biology, we find that 13 has been contaminated by several other symbols and ideas, almost all of them negative. These symbols and ideas are still very much alive, even though we do not consciously always recognize them. They are reinforced by a regular body function that brings inconvenience and sometimes pain to one half of the human race and is now recognized as a problem for the other half too. Thirteen is not unlucky in a practical sense. Its associations make us feel uneasy.

These associations have deep roots in nature, for 13 is the symbol of the moon. In English common law, there are thirteen lunar months of four weeks in every year. They are not true lunar months, for the time elapsing between a particular phase of one moon and the same phase of the next is not 28 days but 29·53059. But the link is clear enough. The words 'moon' and 'month' share a common origin.

We now have to leap back thousands of years to the age when man was a hunter, rather than a farmer. Night is a good time for hunting and a moonlight night best of all. So primitive man was highly conscious of the comings and goings of the moon and eventually came to measure time by it. When man became a herdsman as well as a hunter, the moon became even more important to him, for on dark nights his animals were at greatest risk from predators. Once he settled down as a farmer, however, his situation changed radically. He needed a means of predicting the time for seed sowing and harvest and for measuring the period for which he would have to store fodder. For this, however, the moon was dangerously unreliable.

We can see why when we consider the present-day Yami fishermen who live on the island of Botel-Tobago off Taiwan. Traditionally, their season starts when the flying fish first rise and they predict this by counting twelve moons from the opening of the previous season. But twelve lunar months are eleven days less than a year. Sometimes, this does not matter. They go out in their boats at night with lighted lamps, the flying fish rise and they start fishing. After two years, however, they are twenty-two days in advance and after three years, thirty-three. There comes a time when they arrive *before* the flying fish have started to rise and return empty handed. They do not go out again the next night or even the next week. They wait for the next moon and so get back on schedule, only to slip behind again in subsequent years.[6]

For fishermen, bad timing is an inconvenience. For farmers, it may be a disaster. If the community's stock of seed is planted too soon, it may rot in the ground and starvation may follow. So a more reliable method of fixing on the right time is essential. The Sumerians and Babylonians solved the problem by using the sun as a basis for their calendar but in so doing, made a fundamental error that has bedevilled almost every Western calendar down to our own times. They tried to build a solar year on twelve lunar months, some with twenty-nine days, others with thirty. This was clearly too short, and to compensate they occasionally slipped in a thirteenth month, often at random, but six times in all during a cycle of nineteen years. The Jews, Greeks, and early Romans followed the Babylonians in basing their calendars on both sun and moon cycles. They too had occasionally to slip in a thirteenth month or the equivalent. This thirteenth month was odd, anomalous, eccentric. From the very beginning, 13 was associated with the awkward and the out-of-step.

Moreover, there was a continuing conflict between the majority who thought of the sun as the more important element and a minority who preferred the moon. It became a conflict between religions and led to the male-dominated cultures of the West. It also left vestiges on our own calendar.

In the early days of Christianity, Christians wished to celebrate the birth of Jesus but the Gospels did not reveal the actual date. The Eastern Church settled for 6 January. Now the sun year started on 25 December, the old date of the winter solstice when the sun was reborn. January 6 comes twelve days later, exactly the difference between a solar year of 365 days and a lunar year of 377 days. It seems clear, therefore, that 6 January was a sort of lunar New Year and there is nothing surprising in its adoption by the Eastern Church. For thousands of years, it had been usual for one religion to take over the festivals of a predecessor.

The Western Church had a different problem. Its main rival was Mithraism whose followers worshipped Mithras, god of light and the sun. The only way Christianity could hold its own in the West was to turn a blind eye on the Mithraic beliefs still held by many Christian converts. So the Christian God came to be worshipped on Sun-day, not on the Jewish sabbath, and like the votaries of Mithras, Christians faced east towards the rising sun when worshipping. They also chose 25 December, the date of the sun's rebirth, as the official birthday of Jesus.

An early Christian Syrian writer made this clear:
'It was the custom of the heathen to celebrate on the same twenty-fifth of December the birthday of the Sun, at which they kindle lights in token of festivity. In these solemnities and festivities, Christians also took part. Accordingly, when the doctors of the church perceived that the Christians had a leaning to this festival, they took counsel and resolved that the true Nativity should be solemnised on that day.'[7]

This probably happened early in the fourth century, and in A.D. 375, the Eastern Church also recognized 25 December as the date of the Nativity. Even then, 6 January could not be forgotten and it became the feast of the Epiphany, when the infant Jesus was first shown to the Magi.

The sun religion may have prevailed officially, and in Anglo-Saxon England 25 December marked the beginning of a new year. Yet the moon religion never entirely lost its influence, lingering on in folk customs right down to our own times. Pennethorne Hughes, a former BBC producer, describes two of them. First, there was the custom in Gloucestershire, England, at least until 1841, of lighting twelve small fires and one large one on 6 January.* 'The people assembled, from miles around, to "burn the old witch"—to sing and drink and dance and celebrate.' In 1941, Hughes came upon the Haxey Hood Game which had been played on 6 January in Lincolnshire, England, for hundreds of years. Thirteen took part—eleven 'boggens', a lord and a fool with a wand of thirteen willows. In a BBC broadcast, one of them explained:

> 'It is played with thirteen hoods, twelve sack hoods and the sway hood which is made of leather. Led by the lord, one by one the sack hoods are thrown up into the air and scrambled for by nearly everybody, and if it is touched by a boggen, it's boggened again, that is, returned to play. This goes on for roughly an hour. Then comes the great moment when the sway hood is thrown; this is the signal for the whole crowd to join in. A general cry of "sway" goes up and then the fun begins.'[8]

In each case, we have a pattern of 12 plus 1. Bearing in mind that both ceremonies take place on 6 January, the start of a new lunar year, it seems clear that the 12 stands for the twelve solar months and the one for the extra month in the lunar year. Similar ceremonies taking place on 6 January have been noted in Worcestershire and Herefordshire and the link with the moon confirmed by the placing of a cake on one horn of the best bull or cow in the byre. The horns represent the horns of the crescent moon.

We find this combination of 12 plus 1 in many ritual groupings: Jesus and his twelve disciples, the cells of the heretical Catharists in the thirteenth century, Robin Hood and his foresters, King Arthur and the knights of the round table, the Scandinavian

---

*January 6 is also of course Old Christmas Day, the day on which Christmas was celebrated before eleven days were 'lost' in the calendar reform. Some 6 January customs are clearly linked with the original Christmas. Equally clearly, others are not.

chieftain Rollo and his warriors.* In their time or origins, all these were rebels, heretics, fairies, or pagans. The oldest, commonest, and most persistent of all such groups is, of course, the witch coven. Whether in Britain, France, Germany, or North America, it seems always to have consisted of twelve members and a Grand Devil. Its symbol was traditionally the pentacle surmounted by a triangle, giving a total of thirteen sides.

We need not describe in detail the conduct of a sabbat with its kissing of the Grand Devil's buttocks, its naked dancing round a phallic pillar, the sacrifice of an animal, and finally the sexual orgies in which the Grand Devil served each woman in the coven, a somewhat painful procedure for he used an artificial penis. Here, we need only note some obvious connections with both the moon and fertility rites. Major sabbats nearly always took place between midnight and sunrise at full moon and the Devil himself wore a mask with a candle set between horns. The ritual dancing and the orgies that followed are echoed in fertility rites practised in Europe, America, Africa, and Australasia, both in the distant past and sometimes too in the present.

The odd thing, of course, is why the moon should ever have been linked with fertility rites which would seem more appropriate to the sun. And why should either the moon or fertility rites ever have been regarded as abhorrent in the first place? Most of us now think of the moon in romantic terms and some still believe they will get good luck if they turn over their money in their pockets when they first see the new moon. Others have found in it the origin of hot-cross buns, originally moon-buns, inscribed with two crossing crescents long before the Crucifixion. Even the nursery rhyme Jack and Jill has been traced back to the Norse myth of Hjuki and Bil who were whirled to the moon as they were carrying a pail of water on a pole. Ancient fertility rites can be traced in the throwing of rice at weddings and in dancing round the maypole. All these are pleasant associations, quite different from the horror with which Western man has usually regarded fertility rites and the moon.

To find the reason for this negative attitude we have to go back

*We occasionally find groups of 13 such as judge and jury, mayor and corporation, holding a place in respectable society but most do not.

some 4000 years when Abraham led the Israelites down from Haran in northern Mesopotamia into the land of Canaan. They found there a people who had to live with a highly treacherous climate whose searing summer winds and unreliable autumn rain continually threatened their crops and even their survival. The Canaanites sought help from Ashtaroth, a goddess also known as Asherat, Ashtart, Istar, Ishtar, and later Astarte, who wore on her head a crescent or pair of horns enclosing a disc and who was clearly a moon goddess in origin, although she later became identified with the Morning Star and so with Aphrodite and Venus. She was sometimes the mother of Ba'al, sometimes his wife. The Canaanites wished her to make their land fertile and their worship took the form of a fertility cult. Blood sacrifices were one feature of this but ritual prostitution and sexual orgies were much more usual. To the prophets and priests of the Israelites, Ashtaroth was an abomination, not simply because she was a false goddess but because of the ways in which she was worshipped. The words 'whore' and 'harlot' sprang as readily to their lips as 'imperialist capitalist pig' to the lips of a Chinese communist.

Yet the Israelites, too, faced extinction by drought. At one stage, they were driven to take refuge in Egypt and when they returned in the thirteenth century B.C., they too felt the need to propitiate the nature goddess or, as the Bible puts it, 'to commit whoredom with the daughters of Moab' (Numbers 25.1). Some 700 years later, the Lord still found it necessary to reprimand them through Ezekiel (16.15 ff) for their continued obsession with blood sacrifice and temple prostitution. 'Thou hast taken thy sons and daughters, whom thou hast borne upon me, and these hast thou sacrificed unto him to be devoured . . . Thou hast moreover multiplied thy fornication in the land of Canaan unto Chaldea . . .' So the struggle between the Israelites' God with his rigid moral code and the indigenous moon goddess who demanded worship of a totally different kind is a continuous strand in the history of Israel, and since the Jewish God became the patriarchal God of our own culture, we also inherited their hatred of Ashtaroth. The Christian Trinity has no place for the feminine.

In Greek mythology, too, the moon goddess has horrific aspects. She is Selene or Mene, the indescribably beautiful mistress of

Zeus but also of the demonic Pan. She is Hecate, patroness of sailors, of worldly wealth and, after plunging into the underworld, of sorcery and witchcraft. Hecate, with her pack of ghostly hounds, haunted cross-roads where three ways met and it was the custom to build in these places 'triple Hecates' with three heads, often of boar, hound, and horse, at which offerings were laid at full moon. The moon goddess was also Artemis, virgin huntress who sometimes exacted human sacrifices and became so angry when the huntsman Actaeon saw her naked that she changed him into a stag and watched his own hounds tear and eat him. Yet at Ephesus, she was also worshipped as the goddess of fertility and if we can believe the sculptors, had not two breasts but dozens, ranged in rows across her chest.

Whether Selene, Hecate or Artemis, the moon goddess was almost always the daughter of Zeus and it is clear too that she was a composite of numerous local deities, most of whom presided over nature or fertility. In Roman mythology, most of her attributes were taken on by Diana, who was also worshipped at three-faced statues at cross-roads. At Aricia, Diana's shrine was presided over by a runaway slave who had killed with a living branch the previous priest and was himself doomed to suffer the same fate.

In all these classical manifestations of the moon goddess, therefore, we see a wild, cruel aspect of the feminine, a sexuality that is anything but domesticated, a craving for blood and vengeance which is almost always directed against men.* Hecate's embracing of witchcraft, with its perverse phallic symbol of the broomstick made of unlucky elder (see below) and its 13-sided insignia fits easily into the pattern. So does Shakespeare's comment: 'His mother was a witch, and one so strong/That could control the moon'. He also wrote:

'Therefore the moon, the governess of floods,
Pale in her anger, washes all the air,
That Rheumatick diseases do abound: . . .'

Even today, we find similar prejudices in the school playground. Iona and Peter Opie, the British experts on the culture of

*Some of the Goddess's positive aspects passed into the cult of the Virgin Mary, 'Queen of Heaven'.

childhood, wrote: 'Outwardly, the children in the back streets
and around the housing estates appear to belong to the twentieth
century, but ancient apprehensions, if only half believed in, continue to infiltrate their minds; warning them that moonlight
shining on a person's face when he is asleep will make him go
mad . . .'9

This hatred of the moon and everything linked with it also contaminated the moon number, 13. In the whole of the Bible, the
only reference to 'the thirteenth day', six in all, come in the book
of Esther which tells how Ahasuerus, king of Persia, called
together his scribes on the thirteenth day of the first month and
ordered his officials to kill every Jew in his kingdom on the
thirteenth day of the twelfth month. In the event, the Jews turned
the tables and slaughtered *their* enemies on the fateful thirteenth.
The book of Esther is fictional, so the number is not just a
coincidence of history; nor can we ignore the fact that it is
repeated six times. It was clearly chosen by the unknown author
precisely because of its horrific connotations. Significantly, Esther
is thought to have been written in the second century B.C. after
the Israelites had returned from their exile in Babylon. As we have
seen, the Babylonian calendar was permanently out of joint because
it was impossible to fit thirteen lunar months into a solar year.

From Esther on, we find 13 regularly linked with doom. By
Roman times, it was already thought unlucky for thirteen people
to sit down together at table. In the ancient Celtic tree alphabet,
each consonant was the initial of a tree after which it was named
and R, the thirteenth consonant, stands for the Ruis or elder. Now
elder was always regarded as evil. The Cross was reputedly made
of elder wood. No baby would thrive in an elder cradle, nor
would children beaten with elder switches. In Ireland, witches'
broomsticks were said to be made of elder. According to Robert
Graves, 'So unlucky is the elder that in Langland's *Piers Plowman*
Judas is made to hang himself on an elder tree. Spencer couples
the elder with funeral cypress, and T. Scott writes in his
Philomythie (1616):

> The curs'd elder and the fatal yew
> With witch [rowan] and nightshade in their shadows grew.
> . . . The elder is the tree of doom.'10

This is not to say, of course, that every time a home-owner asks for his street number to be changed from 13 to 12a that he consciously thinks of elder, witchcraft or the moon. Clearly he does not. Yet like so many other prejudices and preferences that influence us without reasonable cause, fear of 13 is embedded deep in our unconscious minds. The number is inextricably tangled with all that is most bestial and demonic in ourselves and in other human beings. We would rather not know about it.

Here the story takes another twist. It may be thought irrational to link the moon and its 13 annual visitations with human behaviour. Yet we are forced to do so more and more in the light of recent discoveries.

It has long been known that the moon rules the earth's tides and hence the life cycle of many forms of marine life but the subtlety of its influence is only now being fully appreciated. The work of Dr Frank A. Brown of Northwestern University is well known, especially his experiments with oysters taken at Long Island Sound on the east coast of America. Moved a thousand miles inland to Evanston, Illinois, and kept in darkened, pressurized tanks, they continued to open and close their valves to the rhythm of the tides at Long Island. After two weeks, however, they gradually changed their rhythm to that of the tides that *would* have been found at Evanston if that city had been on the coast. Though shielded from the light of the moon, they were clearly governed by its movement relative to the earth and subsequent experiments suggested that their body clocks might well be affected by changes in terrestrial magnetism for which the moon was responsible. If oysters are influenced in this way, can the moon have a direct effect on man?[11]

People have always believed so. There is an old Cornish saying 'No moon, no man', which means that a baby born during a period when no part of the moon is visible will not live to be an adult. In north-west Germany, the moon was regarded as a midwife, so that births were said to be more frequent when the tide was rising and in Charles Dickens's novel *David Copperfield* there is a famous passage which hints that the moon also rules our exit from life. As Barkis lies dying, Mr Peggotty tells Copperfield, 'He will pass away with the tide.' When Copperfield asks 'Which tide?', Mr Peggotty goes on: 'Along the coast, people always die at the ebb tide and they're born at the flow. He will go when the tide goes.' Mental illness too was regarded as 'lunacy' because it was believed that victims were 'moonstruck' and liable to uncontrollable outbursts at full moon. The superintendents of mental hospitals used to put staff on overtime when a full moon was expected and even gave inmates a precautionary whipping.

On the whole, doctors and scientists did not follow up these widely held beliefs. Until recently they had no way of explaining how lunar influence might work and they shied away from anything that smacked of astrology. Besides, the germ theory of disease, together with an ever-increasing knowledge of the body's workings, led to spectacular advances in surgery and life-saving drugs. In the treatment of mental illness, professional interest has veered towards behaviourism, chemotherapy, and various forms of depth analysis.

Yet there is now some evidence, though far from watertight, that there may well be some truth in the widely held view that mental hospitals tend to have more admissions at full moon. Patients already admitted may also be more disturbed than at other times. In America, the Philadelphia Police Department reported that cases of fire-raising, kleptomania, homicidal alcoholism, and other crimes against the person increased in number as the moon waxed and decreased as it waned.[12]

It also seems possible that the moon influences the time of birth. A study of more than 11,000 births over a period of six years at the Methodist Hospital of Southern California in Los Angeles showed that six babies were born when the moon was waxing for every five born when it was waning. In case it should be thought that this implies we are just puppets of nature, it is worth pointing out that the pattern was broken whenever the waning period happened to fall nine months after Christmas or New Year celebrations. At these times, the number of births was much higher than expected. In Freiburg, Bavaria, Dr W. Buehler found a similar pattern after studying some 33,000 births, with the added refinement that boys tended to be born on the wax and girls on the wane.[13]

There is a prima facie case for thinking that the moon may affect other bodily processes. In Florida, Dr Edson Andrews who specializes in ear-nose-throat surgery kept records of 'bleeders',

patients who bled so profusely after their operation that they
needed special care. He found that 82 per cent of them had their
operations around the time of the full moon. He concluded:
'These data have become so conclusive and convincing to me, I
threaten to become a witch doctor and operate on dark nights
only, saving the moonlit nights for romance.'[14]

In Chicago, Dr W. F. Petersen claimed to have discovered a
correlation between the phase of the moon and some other conditions. Patients suffering from tuberculosis, for instance, were
most likely to die a week after full moon and least likely on the
eleventh previous day, possibly because of moon-induced changes
in the acidity or alkalinity of their blood. Other workers have
reported links between the phases of the moon and other blood
conditions, not to mention the occurrence of pneumonia and the
time of death.[15]

These findings are not, as yet, generally accepted, but that does
not mean that they need be rejected, only that we must keep an
open mind. If we want a tentative explanation of how the influence might work, we could find it either in the moon's gravitational pull or in the work of Dr Leonard J. Ravitz who correlated
lunar phases with regular changes in the electric field given off by
the human body. Meanwhile, it is a fact that popular belief
associates the moon with death, disturbance, and the female
function of birth.

The moon is associated with another female function which
normally parallels the lunar months and has an extremely important place in the 13 syndrome. It is regarded as almost anything
from a minor inconvenience to a bitter scourge and is still widely
called 'The Curse'. I refer, of course, to menstruation.

The average menstrual cycle is 29 days and occurs some 13
times a year. Its length is so close to that of the lunar month of
29·53 days that the *Concise Oxford Dictionary* equates them, defining
the menses as 'flow of blood from mucous coat of uterus of
female occurring on the average at intervals of lunar month.' But
is it really linked with the moon?

The Swedish chemist Svante Arrhenius (1859–1927) kept
records of 11,807 menstrual periods, compared them with the
phases of the moon and found that they were slightly more likely
to start when the moon was waxing than at other times. The idea

of a link was attractive and even led a German biologist to suggest that it could be traced back to primitive man's tendency to chase women on moonlight nights. The sexual stimulation so produced would start off a regular body process which, he said, eventually became hereditary, giving present-day women a built-in syndromicity with the moon. Unfortunately for this theory and others like it, later surveys of tens of thousands of menstrual periods have produced contradictory results and though it would be premature to insist that there is *no* link, we can say quite definitely that its existence has not yet been proved.

One thing no one can dispute is the dislike and even fear of menstruation throughout the whole of history and in almost every culture. In Leviticus, the Israelites were warned, 'If a woman have an issue, and her issue in her flesh be blood, she shall be put apart seven days: and whosoever toucheth her shall be unclean until the even. And everything that she lieth upon shall be unclean; everything also that she sitteth upon shall be unclean.' (15. 19–20). The Roman naturalist Pliny was more specific. The presence of a menstruating woman would blight crops, sour wine, blunt knives, rust iron, tarnish mirrors, and kill bees. Fruit would fall from the trees and mares abort their foals, especially if the moon was waning. Writing in 1922, Sir James Frazer found similar prejudices widespread in Europe, from Brunswick in North Germany where a menstruating woman at a pig-killing was believed to make the pork putrefy, to Calymnos, Greece, where a menstruating woman in a boat would bring about a storm.[16]

The menstruation taboo is still strong in primitive societies. Among the Arapesh, who live in the Torricelli Mountains of New Guinea, menstruating women withdraw to a crude hut set apart from the village while their husbands look after their families. Among the Bororo, a nomadic people of Niger in Africa, menstruating women are similarly isolated and kept away from crops and animals.

Ignorance and prejudice persist in Western society, though in a somewhat different form. The old magic is discredited; the new gods are hygiene and aesthetics. It is still widely believed that menstruating women endanger their health if they bathe, wash their hair, or eat frozen foods and though the old religious prohibition against sex is rarely heard of, Dr Isabel Hutton could write as recently as 1938 in *The Sex Technique in Marriage*, 'Intercourse should not take place during the period for reasons, chiefly aesthetic, which will be obvious.'[17]

Her advice is still widely followed. According to a survey of 960 couples in California, no less than half ruled out sex completely during menstruation. However 'permissive' our society, few women feel able to discuss menstruation in mixed company, still less to state openly that they are 'having a period', and go to extreme lengths to hide the fact. Advertisements for tampons and sanitary towels stress the need to avoid 'embarrassment', and novels and films, which openly portray almost every type of hetero- and homosexual practice, rarely even hint at menstruation.

Why is the taboo almost universal? Why should almost every society regard with suspicion or disgust a bodily function as natural as defecating or blowing our nose?

Whether or not it is justified, I suggest the real reason is that inherited attitudes are continuously reinforced by the change that so many men see in their own womenfolk thirteen times a year. They may have no particular feelings about menstruation but they can hardly ignore the turmoil that so often precedes it. The group of symptoms known as premenstrual tension can turn a usually serene wife or mother into an irritable, bitchy, even violent shrew.

Various surveys have estimated the proportion of women affected at anything between 12 and 90 per cent, but there is fairly general agreement that between 60 and 75 per cent suffer at least some of the symptoms. According to some researchers, those assessed as 'neurotic', 'socially immature', or 'feminine' have more than their fair share. This may mean only that such labels are pinned on women *because* of the suffering they happen to undergo. Certainly, it now seems unlikely that cultural attitudes have as much effect as was once thought. The Los Angeles psychiatrist Dr Oscar Janiger and a number of colleagues sent a detailed questionnaire to American, American Indian, Greek, Japanese, Nigerian, and Turkish women asking not just for their premenstrual symptoms but also for their views on sex, religion, discipline, and family life. All groups reported a similar pattern of symptoms. Zoo-keepers reported that chimpanzees, gorillas, and rhesus monkeys also suffered.[18]

Dr Katherin Dalton, the British doctor who pioneered research on the premenstrual syndrome, has shown that its effects are unexpectedly widespread. Their influence is felt through the whole of society. A survey of women in Holloway Prison, London revealed that 63 per cent were serving sentences for crimes committed during the premenstrual or menstrual period. In a French study, the proportion was 84 per cent, and in New York almost two-thirds of the violent crimes for which women were brought to book were committed at this time.

The cost to industry is staggering. It has been estimated that British firms lose 120 million working days a year because of menstrual problems, American firms some five billion dollars in lost production. Women are often slower, clumsier, less accurate, more prone to aches and pains and above all more liable to accidents during the premenstrual period. They are also more liable to have accidents at home and on the roads, and a survey of women in the accident wards of four London hospitals showed that half had come to grief at this time. In girls' schools too we find a similar pattern of absenteeism, rebellion against school rules, and work well below the usual standard. This is a serious disadvantage if examinations have to be taken.[19]

Do men also have monthly changes of mood? Some researchers claim to have found evidence that they do so. Others have quoted studies in mixed schools and records of eye-fluid pressure to show that they do not. Any cases that do arise, they say, can be traced to the wife's premenstrual symptoms and disappear when the wife's problem has been solved.

It seems likely that further research on the monthly cycles of both men and women will uncover physical and psychological effects of which we have as yet little inkling. Meanwhile, we are left with menstruation and the period of turmoil that so often precedes it, when so many women (for whatever reason) tend to be 'not themselves', when, like Artemis, they loose their hounds or like Hecate cast a witch-like spell on all around them. This is the dark side of the moon, which attracts so many negative projections and contaminates so many of the symbols associated with it, especially the number 13.

5

Odds, Chance, and Luck

NUMBERS are like dreams. If we know how to read them, they may give us clues to the present, past, and even future. For compulsive gamblers, they hold the key to doom or fortune.

No one understood this better than Joseph Conrad. In his long short story *The End of the Tether*, he created the character of George Massy, a ship's engineer, who won 'the second great prize in the Manilla lottery', bought his own ship with the proceeds and became obsessed with the idea of winning again:

'With his elbows propped, his head between his hands, he seemed to lose himself in the study of an abstruse problem in mathematics. It was the list of winning numbers from the last drawing of the great lottery which had been the one inspiring fact of so many years of his existence . . . There was in them, as in the experience of life, the fascination of hope, the excitement of a half-penetrated mystery, the longing of a half-satisfied desire.

'For days together on a trip, he would shut himself up in his berth with them . . . and he would weary his brain poring over the rows of disconnected figures, bewildering by their senseless sequence, resembling the hazards of destiny itself. He nourished a conviction that there must be some logic lurking somewhere in the results of chance. He thought he had seen its very form . . . Nine, nine, nought, four, two. He made a note. The next winning number of the great prize was forty-seven thousand and five. These numbers of course would have to be avoided in the future when writing to Manilla for the tickets. He mumbled, pencil in hand . . . "and five. Hm . . . Hm." He wetted his finger: the papers rustled. Ha! But what's this? Three years ago, in the September drawing, it was the number nine, nought, four, two that took the first prize. Most remarkable. There was a hint there of a definite rule! He was afraid of missing some recondite principle in the overwhelming wealth of his material. What could it be? and for half an hour he would remain dead still, bent low over his desk, without twitching a muscle.'[1]

Massy was right in principle. By studying the numbers involved, we can accumulate huge sums by gambling. But we have to go about it the right way. To succeed, we have to choose the right game, the right side, and the right set of numbers, and we need to know what we can expect of them. That's how gambling has become a worldwide industry providing many thousands with their weekly pay cheque and pouring a Niagara of cash into the coffers of governments, charities, and the private enterprise promoters of football pools, numbers games, casinos, slot machines, bookmaking, and bingo. Nearly everyone takes a cut of the proceeds. If we do not own shares in the industrial groups that have diversified into gambling, the odds are that we have an indirect interest through the investments of unit trusts, mutual funds, pension schemes and life assurance, not to mention the contribution that gambling makes to the national income through taxation.

Nor does all the profit find its way into the tills of the promoters and tax gatherers. In January 1963, the British actor Sean Connery backed number 17 three times in succession in a roulette game at the St Vincent casino in Italy. By a 50,652 to 1 chance, the marble rolled into the seventeen slot each time and Mr Connery collected some £10,000.[2] In January 1973, Mrs Doris Binfield of Findon in Sussex, England, threw a handful of numbered counters on the floor, picked up eight and used them as her selections in a football coupon. She staked 60p and won a half-share in what was then a world record payout of £546,239.[3]

Both Mr Connery and Mrs Binefield were 'lucky', a concept we shall examine later, but serious gamblers have shown that luck is not essential. A few have spotted situations in which they could not fail to make a killing. Others have won consistently over many years by systematically improving the odds in their favour. This requires no special insight or aptitude but only an understanding of the way numbers work in gambling. The main reason why this is not more widely appreciated is that it often seems to fly in the face of 'commonsense'.

Much of what we know was discovered in the seventeenth and eighteenth centuries when scholars were less cloistered than now and mixed socially with wealthy rakes who picked their brains for advice on gambling problems. So it was with Galileo (1564–1642), the Italian mathematician, astronomer, and physicist who constructed the first astronomical telescope, confirmed Copernicus's theory that the earth and other planets moved round the sun and fought a losing battle with the Inquisition over his views. Between whiles, he found an answer to a question posed by a gambling friend: 'Why, when I throw three dice, do the pips more often total 10 than 9?' After a brief study of the problem, Galileo was able to show that three dice could produce 216 different combinations and of those 27 gave a total of ten pips but only 25 gave 9.

It was an isolated experiment but it seemed to show that *in the long run* a gambler who consistently put his money on 10 in a three-dice game would win out over an opponent who consistently backed 9. But how exactly could he measure the advantage? How certain was it? And what was 'the long run'?

These questions greatly interested Hieronimo Cardano (1501–76), a mathematician and astrologer of Milan, Italy. Usually known as Cardan, he is famous for his solution of cubic equations in algebra, a solution he is now known to have stolen from a fellow mathematician, Nicolo of Brescia. The theft was wholly in character, for Cardan was a violent, unpredictable man who cut off his own son's ear in a fit of temper and was so arrogant that he predicted the date of his own death. When the time came and it seemed unlikely that he would die, he killed himself to prove his infallibility.[4]

We first hear of Cardan as a student at Padua where he gambled to pay his expenses and used his mathematical talents to give him an edge over his opponents. He started from first principles. A die had six sides and it could fall with any one of them uppermost. He called each throw a 'case' and if the required number came up, it was a 'favourable case'. The probability of any particular number coming up with a single throw of one die was one case in six, giving the formula:

$$p \text{ (or probability)} = \frac{f}{c} \text{ or } \frac{\text{(the number of favourable cases)}}{\text{(the number of possible cases)}}$$

So the probability of a head coming up when we toss a coin is one favourable case divided by two possible cases or $\frac{1}{2}$, which we can also express as 0·5. The same is true of tails. If we toss a double-headed coin, however, and call 'heads', the number of favourable cases is two, the same as the number of possible cases. The probability of throwing a head is therefore $\frac{2}{2}$ or 1, the highest degree of probability, in other words certainty. If we called 'tails' on our double-headed coin, we should have two possible cases but no favourable cases or $\frac{0}{2}$, which is 0. This is the lowest degree of probability and means that an event is impossible.

Putting it in words, the Theory of Probability runs: *The probability of an event is equal to the number of favourable cases divided by the number of possible cases, provided that all the possible cases are equally likely to occur.*

The proviso is important. Dice can be loaded, bevelled, edged, and otherwise tampered with to ensure that one or more sides will come up more often than others. Even a genuine coin in mint condition may be designed on such a way that aerodynamically one side has a slight advantage. Both dice and coins can be rolled or tossed in a manner that influences the result and there is even some evidence that a few people may be gifted with the power of psychokinesis, the ability to change an object's motion just by thinking about it. The Law of Probability excludes all these. It assumes a fair die or coin, a fair roll or toss, a complete absence of psychic powers, and players who do not peek or deal from the bottom of the deck.

Bookmakers, pools proprietors, and casino owners know all about the Law of Probability and take it into account when they fix the odds.

Now, *correct* odds are the ratio between favourable and un-favourable cases. When we throw a die with the intention of getting one particular number, there is one favourable case (the number we want) and five unfavourable cases (all the rest). So the odds against success are 5 to 1. If we want either of any two specified numbers, there are two favourable cases and four unfavourable, giving odds of 4 to 2 against, which are the same as 2 to 1 against. Suppose we want any one of three specified numbers. The odds now are 3 to 3 or 'evens'. Calling for any number *except* one that we specify gives us five favourable cases and only one unfavourable. The odds are now 5 to 1 *on*.

Hence, the odds against picking out a specified card from a pack of fifty-two are 51 to 1; the odds when calling 'head' or 'tail' at the toss of a coin are evens; and the odds when picking a red counter from a lucky-dip bag containing four red counters and one black are 4 to 1 *on*.

It must be stressed again that these are the *correct* odds and not the odds offered when we place bets commercially. It is easy to see why. Let us take a horse-race with ten runners and assume that each has an equal chance of winning. The correct odds against any particular horse are 9 to 1 but any bookie who offered these odds would be out of business faster than a Derby winner passing the post. If the total money staked was split evenly among the ten horses, he would collect nine units from the losers and pay out nine units to the winner, who would also receive back the unit he originally staked. This would leave the bookmaker with nothing for expenses, let alone profit. He is forced, therefore, to offer less favourable odds.

Could he be sure of coming out on the right side by offering odds as low as 6 to 1 on every horse? At first sight, it might seem that he would still collect nine units from the losers and since he would pay out only six to the winner, he would have a surplus of three units. In practice, though, the money staked is never split evenly among the runners since punters always believe that some horses have a better chance of winning than others because of their form or the skill of their jockeys. In nearly every race, one horse is backed much more heavily than the rest and is known as the 'favourite'. Others attract very little money indeed and are called 'outsiders'. Between favourite and outsiders are horses with varying degrees of support. Now suppose that a total of x units is staked on a ten-horse race and half of this is laid on the favourite. If the bookmaker is offering odds of 6 to 1 against all the runners and the favourite wins, he collects a total of $\frac{1}{2}x$ from the losers but has to pay out $6 \times \frac{1}{2}x = 3x$ to those who backed the favourite. He loses $2\frac{1}{2}x$ and is once more on the losing end.

He therefore aims to 'make a book' which will bring him a profit whichever horse wins. He does this by offering shorter odds on the favourite and other fancied horses, longer odds on out-siders. Favourites commonly run at 5 to 4 but odds *on* a heavily supported runner have been as short as 100 to 1, meaning that the punter has to risk 100 units in the hope of winning 1. Outsiders may start at 20 to 1 against and on these, one unit is risked for a possible gain of 20. A book on a four-horse race which is bound to make a profit for the bookmaker might look like this:

Runner	Odds	Stakes held	Total	To pay if wins	Profit
Horse A	4 to 5 *on*	£50		£90	£10
Horse B	2 to 1	£22½	£100	£67½	£32½
Horse C	2 to 1	£22½		£67½	£32½
Horse D	15 to 1	£5		£80	£20

(Unless otherwise stated, odds are *against*. Payments made by the bookmaker include the original stakes which are, of course, returned automatically to winners.)[5]

This grossly oversimplifies a highly complex series of trans-actions. Nor does it take into account betting through the Totalisator (Tote) or *pari mutuel* where all stakes are pooled, a fixed percentage deducted by the organizers and the rest shared between the winners of various types of bet in a predetermined ratio. However, it does show broadly how odds are worked out for events where factors other than chance are taken into account by backers. These include dog races, cockroach races, beauty contests, and political elections.

Even in games of pure chance, the odds offered commercially must be less than the true odds if the promoters are to stay in business. Roulette, for instance, is an ever popular game whose origins have been traced back to ancient Greece where bored warriors whiled away the time between battles by whirling a shield on the tip of a spear. In its present form, dating back to eighteenth-century Europe, it is played on a dished wheel with an outer rim or back-track and a number of slots round a central boss. A croupier, employed by the house, calls *faites vos jeux* ('make your bets'), spins the wheel in an anti-clockwise direction with his left hand and throws an ivory ball clockwise round the back-track with his right. The ball finally comes to rest in one of the slots.[6]

Two types of wheel are in common use. The French (figure 1) has slots numbered 1 to 36 and also a single zero (o) and is almost universal in Europe. The American wheel also has slots numbered 1 to 36 but it has two zeros (o and oo), making 38 slots in all against the French wheel's 37. This has an important bearing on the odds. Both wheels have alternate black and red slots, except for the zeros which are green, but because of the extra zero, the sequence of numbers is different on the American wheel.

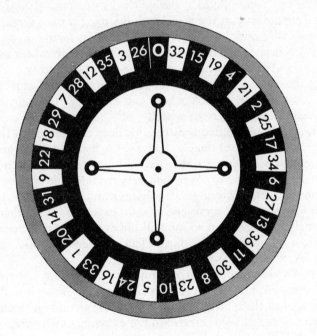

FIGURE 1 French roulette wheel

Bets are laid by placing chips (tokens representing money) on the appropriate section of the table until the croupier calls *rien ne va plus* ('no more bets') soon after he has thrown the ball into the wheel. When it comes to rest, he rakes in losing bets and pays out winners at odds laid down by the house. Basically, players bet on which slot the ball will drop in but they can do this in a number of different ways. These are essentially the same in both Europe and America, though the tables on which players place their bets have somewhat different layouts (Figures 2 and 3). Perfidious Albion often lives up to its reputation by combining a French wheel with an American table.

Among the bets available are *en plein* ('straight'), a gamble that a particular number will come up at casino odds of 35 to 1, and *douzaine* ('dozen') which covers a group of twelve numbers, either 1–12, 13–24, or 25–36 at odds of 2 to 1. There are several even-money bets in which the punter backs *rouge* ('red'), or *noir* ('black'), *pair* ('even') or *impair* ('odd'), *manque* ('1–18') or *passe* ('19–36'). When we remember that the wheel includes one (or two) zeros, it will be seen that none of these are correct odds, for the *en plein* ('straight') bets should pay out 36 to 1 on the French wheel and 37 to 1 on the American wheel and not the 35 to 1 offered on both. Overall, the casino advantage ranges from $1\frac{13}{17}$ per cent to $2\frac{26}{37}$ per cent on the French wheel and from $5\frac{5}{19}$ per cent to $7\frac{17}{19}$ per cent on the American wheel.

Again, it must be emphasized that this does not mean that the casinos are cheating. Presumably, they exist to meet a demand and if they are to stay in business, they need a big enough margin not just to make a profit but to cover their very considerable expenses. The punter expects glittering surroundings and sometimes even free floor shows and free meals and he must expect to pay for them by eventually losing. Even so, he still hopes to leave with more money than he brought with him. Hence his determination to beat the casino at what is literally its own game.

Punters who bet at random sometimes win huge sums but collectively they will probably lose because the odds are so heavily weighted against them. To overcome the house advantage, many use systems, some of them bizarre in the extreme. One painted a matchbox half black and half red, placed a spider in it and at each turn of the wheel bet black or red according to where the spider

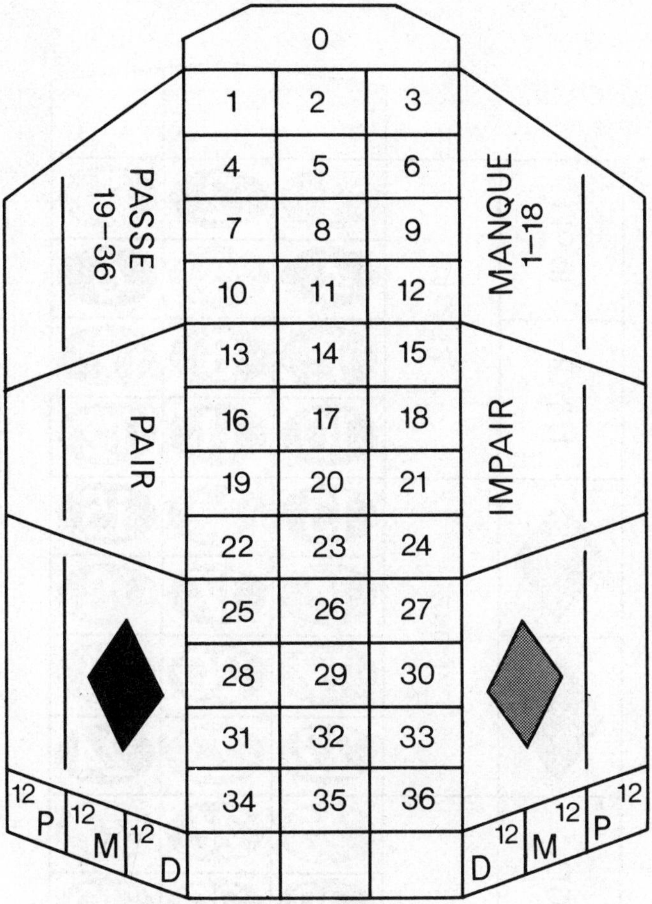

FIGURE 2 French table

happened to be sitting. In the nineteenth century, a French textile manufacturer called Stefan Heller used to work out numbers he thought likely to win by an abstruse analysis of the totals of bobbins employed at his factories on different days. He even travelled with a mathematical secretary to help with his calculations.[7]

Most of us would agree that systems of this kind are as likely to make our fortunes as backing the serial number of our air ticket,

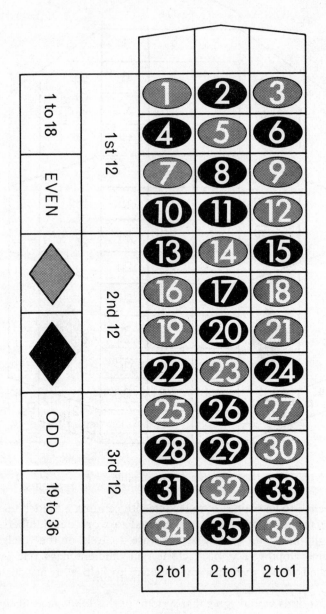

FIGURE 3 American table

the digits of our postal code or the date on which our great grandmother celebrated her golden wedding anniversary. How far can we improve our chances by relying on systems based on mathematics?

Systems fall largely into three categories, the first consisting of those devised by the totally innumerate. Typical of these is one which claims that we can make a steady profit by putting one chip each on *noir, pair,* and *passe* at each spin of the wheel. The idea, presumably, is that we are almost bound to win because 32 of the 36 numbers are covered and many of them more than once. But what of the chips we lose? It takes only a couple of minutes to work out that on both French and American wheels only four numbers would be covered three times, 14 twice, 14 once and 4 not at all. On each number covered three times, we make a gain of 3, on each covered twice we make a gain of 1, on each covered once we lose 1 and on each that is not covered at all we lose 3. Moreover, we must take the zeros into account, raising the number of uncovered slots to 5 on the French wheel and 6 on the American. On 37 spins (French) or 38 (American), the chances are:

Gains.

Four numbers covered three times	12
Fourteen numbers covered twice	14
Total	26

Losses.

Four numbers not covered at all	12
Fourteen numbers covered once	14
One zero (two if American)	3 (6)
Total	29 (32)

Any punter who takes the trouble to do his homework will see that he is more likely to lose than win.

Personally I would find it more exciting to cuddle up in bed, if only with a good book and a cup of cocoa, than play the *noir-pair-passe* system. The ploy of continually backing one number *en plein* and if it comes up letting our winnings ride in the hope that it will come up three times in succession at least promises suspense and high drama. But we cannot hope to make a profit in the long run for once again, the odds are against us. The pay-off of $35 \times 35 \times 35 = 42,875$ is well below the correct odds of 50,652 to 1 (French) or 54,871 to 1 (American). In this case, though, the casino advantage is almost irrelevant, for the chances of our winning are extremely remote and only a fool would embark on such a 'system' in the hope of making a profit. If we do win, our gain is so huge even from a £1 bet that we shall scarcely be interested in what we 'ought' to have won.

The second type of system is based on a fallacy, but one which seems on the face of it commonsense. Certainly it convinced the greenhorn midshipman in Captain Marryat's *Peter Simple* when he asked one of his shipmates how he had survived so many battles. 'He replied that he always made it a rule, upon the first cannon-ball coming through the ship's side, to put his head into the hole which it had made; as, by a calculation made by Professor Inman, the odds were 32,647, and some decimals to boot, that another ball would not come in at the same hole. That's what I never should have thought of.'[8]

Numberless gamblers have thought of it and have even dignified it with the title of the 'maturity of chances' theory. They take the view that if red (or heads or a double-six) comes up five times running, the odds against it coming up a sixth time are very long indeed. Conversely, if red (or heads or a double-six) has not appeared for a considerable time, the chances of it doing so in the immediate future are good and will improve. If asked to justify their view, they would probably say, 'because of the law of averages'.

The snag is that neither cannon balls, roulette marbles, coins nor dice have ever heard of the 'law of averages'. Cussedly, they go on behaving on a basis of pure chance. Even if we had a run of a hundred reds, heads or double-sixes, the chances of another one coming up next are precisely what they always were: 1 in 37 (or 38), 1 in 2 and 1 in 36 respectively. The same rule applies to the midshipman's cannon-ball. The odds may, or may not, be '32,647 and some decimals to boot' that a later ball will not enter by the same hole as an earlier ball, but the odds against a later ball striking *any* specified place are precisely the same.

'Ah', you may say, 'but what about "the long run"? If a coin is tossed enough times, we know that heads and tails even out. If tails fall behind in the reckoning, there must come a time when we get extra tails to catch up. So heads are that much less likely.'

Again, this is not so. There is no law of averages, only a Law of Large Numbers which states that the proportion of heads (or any other specified result) is likely to come steadily closer to the probability of a similar result in one particular case, the longer the run is continued. In other words, the more coins we toss, the more likely it is that the proportion of heads will get steadily closer to the probability of a head at a single toss, that is $\frac{1}{2}$. The word 'likely' is the key. We are dealing in probabilities, not certainties, and we are talking about proportions, not numbers. Even after a million throws, there may well be a large deficit of tails (or heads). Besides, we never know at what stage we are joining the game. If a roulette table has been in use continuously for several years, it may have accumulated a huge deficit (or credit) or *rouges, noirs, pairs, impairs, passes* or *manques*. We simply do not know. So even if the Law of Large Numbers were an exact guide, which it isn't, it could not help us in practice.

Suppose we ignore the past. Is there any way in which we can work out the probability of pairs, threes and longer runs of a particular result coming up during a particular period of play? If so, can we use the knowledge to our advantage?

In coin-tossing, there are only two possible outcomes and we can apply the formula

$$p = \tfrac{1}{2}^{n+2}$$

where p equals probability and n the number of similar results in a run. So the probability of a pair of heads is $\frac{1}{2}^{2+2} = \frac{1}{2} \times \frac{1}{2} \times \frac{1}{2} \times \frac{1}{2} = 1/16$ and the probability of three heads in succession is $\frac{1}{2}^{3+2} = 1/32$ and so on.* With 8,192 tosses, therefore, the runs of heads on the following page are probable:

---

*It must be remembered that we are talking about the probability of two heads coming up in a long continuous run. This implies that the pair of heads must be preceded and followed by at least one tail. So the probability is different from that of two heads coming up in two isolated throws. Here it would be $\frac{1}{2} \times \frac{1}{2} = \frac{1}{4}$.

Number of consecutive heads in a run	Probable number of runs
1	1024
2	512
3	256
4	128
5	64
6	32
7	16
8	8
9	4
10	2
11	1
11+	1

Using the same formula, we can expect a similar number of runs of tails; for the combined number of runs therefore these totals can be doubled. So it follows that in 8,192 tosses, we can expect a run of three heads (or three tails) to be followed by at least one more head (or tail) in all those cases where runs of four or more are indicated. Taking both heads and tails, these total 512. In other words, if we were to gamble on three tails being followed by a head or vice versa, we should expect to win in 512 cases and lose in 512 cases, which is exactly what we should expect by the Law of Probability. No matter how long the run of heads (or tails), the probability of another head (or tail) turning up next is always $\frac{1}{2}$.[9]

Now this has important consequences for roulette players. They are attracted to the even-money bets because these offer the most favourable odds. Yet it is impossible to beat them (except by luck) because a run of three, five or even a million *rouges*, *pairs* or *manques* does not affect the chances of a *noir*, *impair* or *passe* coming up next. Those chances are always $\frac{1}{2}$ *less the advantage given to the house by the zero and double-zero.*

Hence the probable disappointment awaiting the player of the Biarritz system which is built four-square on the quicksands of the 'maturity-of-chances' fallacy. After noting the winning numbers in 111 ($= 3 \times 37$) consecutive spins of the wheel (114 American), the punter backs any that has not already come up ten times with an *en plein* bet of a single unit for the 34 spins that follow. He reasons that it *must* come up at least once during this period, blithely ignoring the Theory of Probability and also the fact that his chance of a 35–1 profit applies only to his first bet. After that, mounting losses have to be set against the eventual gain which disappears entirely after the thirty-fourth failure, leaving him with a loss of 34 units on the run.[10]

Systems relying on the 'maturity-of-chances' theory are not confined to roulette. Nor are those of the third main type, known in casino as 'the martingale' but familiar to small-time gamblers the world over as 'doubling-up'. It offers a fair chance of making a small profit and a remote one of bankrupting the casino. On the face of it, the chances of success seem undeniable if we are patient and not too greedy. We start by backing an even-money chance with a single chip and if successful, end the first series of games immediately with a profit of one chip. If we lose, we back the same even-money chance with two chips, finishing the series with a profit of one chip if we win or doubling up to 4 if we lose. The idea is that we must win eventually and if we continue doubling-up after a loss, we shall end each series with a gain of one chip. Thus, a win delayed until the eighth spin of the wheel would bring losses of $1 + 2 + 4 + 8 + 16 + 32 + 64 = 127$ chips but we should pick up 128 on our next bet, leaving us one ahead of the house.

Apart from being a singularly dull way of passing an evening, the martingale is bound to fail eventually. Even if he has the capital and the nerve to keep on playing through a losing streak, a player will eventually bump up against the house limit, usually 500 times the minimum stake. After losing nine times in a row, he will be down by $1 + 2 + 4 + 8 + 16 + 32 + 64 + 128 + 256 = 511$ units and to hope for a profit, his next bet would have to be 512 units. With a house limit of 500, this would not be accepted and he has lost 511 units on the series. To recoup them, he has to win the next 511 series, yet according to the Law of Probability, he can expect another losing sequence in the next 500 series (300 with an American wheel), putting him still further behind. Of course, this need not happen and it is said that in three cliff-hanging days in 1891, Charles Deville Wells, the original 'man who broke the bank at Monte Carlo', used the martingale system to parlay £400 into £40,000, But he was just lucky. On the basis of probability, the punter for whom the martingale becomes an obsession will, like Sir Francis Clavering in Thackeray's *Pendennis*, break not the bank but himself.

Even if there were no house limit, the martingale would still lose money for us in the long run. An unbroken succession of 28 even numbers has been recorded at Monte Carlo and anyone betting on odd with a basic stake of only £1 would need capital of £536,870,912 to stay in the game. Even if the next number is odd (and there is a 50–50 chance that it will be yet another even), the profit on the series is still only £1. Since the stake has to be available in cash, the punter's gain is far less than he could have obtained by investing his capital overnight through ordinary banking channels. If we cost in his time, not to mention the risk, his loss on such a theoretically winning series is substantial.[11]

At first sight, the reverse martingale offers a better prospect. In this system, the player lets winning bets ride, using the bank's capital to double up, not his own. If he loses, he goes back to his single unit bet. Before embarking on a series of games, he makes up his mind that he will quit after a specified sequence of wins, perhaps the maximum allowed by the house limit. Obviously, he has a very good chance of making comparatively small losses and a remote chance of hitting the jackpot but in the long run, he will be broken by the house advantage.

And so we could go on, for there are almost as many 'infallible' systems at roulette as there are punters to believe in them. Yet there is only one that proved truly infallible. It was devised by William Jaggers, a British engineer, and it earned him a profit of 1,500,000 francs at the end of the nineteenth century. He hired six men to jot down the winning numbers for a whole month and when he went through them, he knew immediately that he had found a way of beating the house advantage. For some of them were coming up much more frequently than the Law of Probability said they should. He concentrated his bets on these and won consistently. For, as he had suspected, the tables were biassed, not because the owners wished to cheat, but because they had allowed the balance to get out of true. The knowledge gave Mr Jaggers an unbeatable advantage but anyone thinking of trying his system today would do well to forget it. Tables are checked daily and any bias corrected before play starts.[12]

With roulette, it is fairly easy to work out the correct odds because each bet is separate and depends on the ball falling into a single slot. Craps, the most popular dice game in America and now widely played in Europe, involves two dice and bets are often made on a sequence of rolls. Few punters are able to work out the house advantage or even to appreciate that it has one.

A 'shooter' opens the game with a 'come-out', that is, he throws two dice so that they bounce off a backboard. If the pips total 7 or 11 ('naturals'), he wins ('passes'); if 2, 3, or 12 ('craps'), he loses ('misses out'). In either case, that particular game is over and a new one starts. A come-out of 4, 5, 6, 8, 9, or 10, however, is known as a 'point' and the shooter keeps on throwing. If he repeats his point number before throwing a 7, he wins. If a 7 comes up first, he loses.[13]

Punters bet on whether the shooter will pass, miss out or throw a particular number. As in roulette, there is an elaborate table layout on which they place their stakes (Figure 4), the odds offered varying with the type of bet made. How favourable these are can be calculated from Figure 5 which shows the 36 different ways in which two dice can fall to produce the various totals from 2 to 12. It will be seen that the chances of throwing point 4 are 3 out of 36 or 1 in 12. The chances of repeating it are again 1 in 12 but the chances of a 7 coming up are 1 in 6, so the probability is that the shooter will miss out. In 1,980 games, a point 4 come-out will probably be thrown 165 times but the shooter will go on to win only 55 times. Similarly, we can work out the probability of all possible results, assess the correct odds and compare them with those offered.

Let us first take the 'pass' or 'front line' bets, meaning that we are backing the shooter to win. The bank usually offers 'evens', one unit for every unit staked, but the true odds are better than evens. In 2,029 games, the shooter 'should' win only 1,000 times and a punter staking one unit on each game with correct odds of 1,029 to 1,000 would lose 1,029 units and gain 1,029 units so that he ended up with his original capital intact. At evens, he loses 1,029 units but gains only 1,000, giving the house a profit of

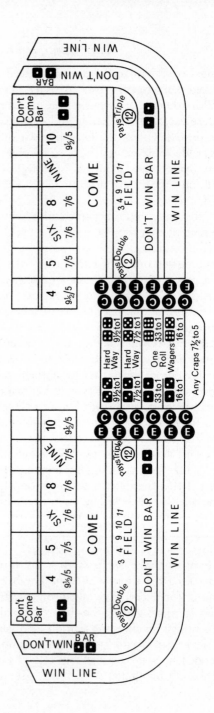

FIGURE 4
Layout of a
craps table

29 units on the total of 2,029 units he has staked. This is slightly more than 1·414 per cent.

If the punter backs the shooter to lose with a 'don't pass' or 'back line' bet, the bank again offers evens. Since the shooter is more likely to lose than win, making the correct odds less than evens, this would seem to be a guarantee of long-run profit to the punter. However, a come-out of double-6 is usually barred and if the shooter throws double-6, it does not count. 'Don't pass' bets are allowed to ride for the next come-out but the restriction sharply reduces the shooter's chances of losing and he is, in fact, more likely to win. So the true odds for the punter who has backed him to lose are better, not worse, than evens and once more the bank has a long-run advantage, this time of 1·403 per cent. If, as sometimes happens, double-1 is barred instead of double-6, the bank's advantage is the same but a bar on 1, 2, which is sometimes imposed, excludes not one but two losing come-outs and the bank's advantage on 'don't pass' bets soars to 4·385 per cent.

Other plays range from the straightforward 'action' bet that a particular combination will be thrown next, to bets on more elaborate sequences or combinations. On some of these, the bank's advantage is as much as 27 per cent. Most favourable to the punter is the so-called 'free' bet sometimes allowed to 'pass' bettors when the shooter comes out on a point. They may double their stake and if they win, the second half of their bet is paid at correct odds. The bank exacts its usual advantage on the first half but overall, its profit is well under one per cent. Obviously, it pays punters to take all the 'free' bets offered and to stick to other bets where the house advantage is low. Apart from 'pass' and 'don't pass', these are mainly 'come' and 'don't come'. Most of the rest give the bank an advantage of at least 5 per cent. In the long run, there is no known way of beating this advantage and the innumerable systems for which claims are made are based on faulty arithmetic, come to grief on the maturity-of-chances fallacy or fall foul of the house limit.

How, then, do we account for punters who gain consistently more than they lose when they bet commercially, even though they play only games of pure chance where skill is not involved? Whether or not they use a system, we cannot explain their success

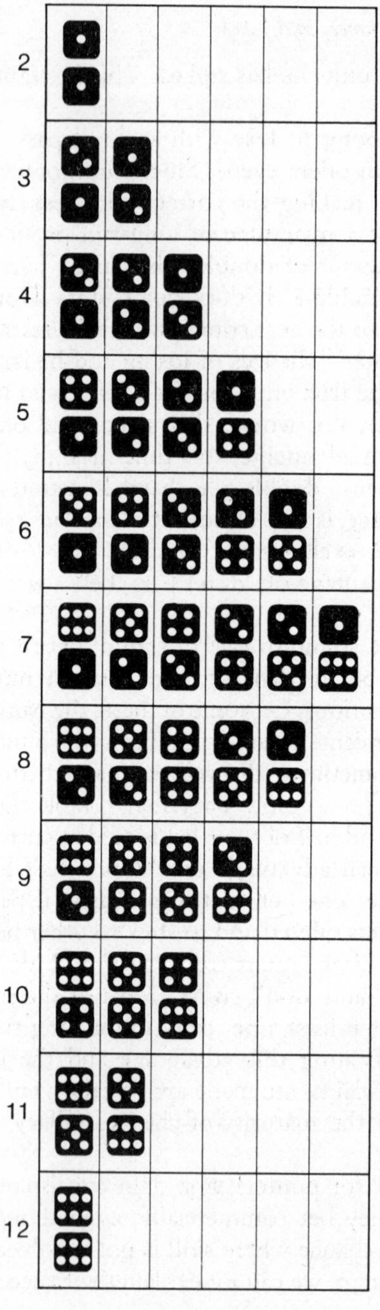

FIGURE 5
Thirty-six ways
in which two
dice can fall

by the way they play, for even the shrewdest appreciation of odds would enable them only to minimize losses, not make a profit. Clearly, we must look for some other factor.

One possibility was suggested by Dr Julian Tynner, a former lecturer at the London School of Economics, who was in charge of Britain's psychological interrogation unit during the Second World War. Over a period of eight weeks, he studied 1,000 regular punters at Monte Carlo, half of them men who gambled alone or with other men, the other half men who were accompanied by their wives or girl friends. Dr Tynner found that those with a wife or girl friend were consistently more successful than those who gambled alone or with other men and he found a similar pattern at several other gambling centres, notably Hamburg, West Berlin, London, and Manchester. He thought it might be explained by the 'pre-cognition gifts which are a strong part of the intuitive make-up of at least 90 per cent of women.' Winning numbers came into their minds spontaneously *before* the croupier spun the wheel. 'Sometimes a wife will tell her husband what numbers to play; in other cases she will merely glance at him and transmit it by thought transference . . . It's no coincidence that some 60 per cent of pools winners are men who have played hunches on the basis of numbers supplied to them by their wives.'[14]

There is now a good deal of evidence in favour of telepathy and precognition[15] but not enough to convince everybody. When we try to account for success in gambling, the known working of chance itself offers a much simpler explanation.

We know that in a very large number of tosses, a coin will probably work closer and closer to an equality of heads and tails but no one has suggested that we are likely to approach this point in the number of tosses any of us is likely to make in a single lifetime. The probability is that there will be a substantial inequality and if we bet a similar amount on each toss, we shall end up richer or poorer than when we started. If it happens to be richer, we count ourselves 'lucky'.

In coin-tossing, of course, we should be betting on correct odds, which are evens, but in commercial betting too, when the house has an advantage, it expects to make a profit only in the long run from thousands of customers and among these, there will be some who beat the house advantage by pure chance. We call them 'lucky', but the term is meaningless. Events are taking one of a very large number of equally likely courses.

Similarly with cards: it always rates a paragraph in the newspapers if someone picks up a bridge hand and finds he has a complete suit of spades. The odds against this happening are 635,013,559,599 to 1 and we are duly astonished. Yet the odds against any specified hand being dealt are exactly the same. They surprise us only when some pattern in the hand demands our attention, and the pattern that is most striking is one that gives a certain chance of winning.

In any event, or series of events, it is almost certain that chance will favour some person, or persons, more than others. Many years ago, a friend of mine made the point graphically in a short story that later became a television play. As I remember it, a bank cashier received a racing tip by post from an anonymous tipster using an accommodation address. There was no fee, just a request for a proportion of the winnings. The cashier was not a betting man and knew nothing of the turf but staked a small sum and to his surprise, won. He sent the tipster his share and had almost forgotten the incident when a second tip arrived from the same source. He backed this too and won again. As further tips came, he raised his stakes and kept on winning. A frightening thought occurred to him. He had access to large sums of money at the bank. Suppose he borrowed £100,000 to make one final bet. If he won, he could replace it without being detected and retire in luxury. If he lost, he faced disgrace and prison. The first six tips had been successful. So the tipster clearly had a winning system. Was it worth the risk?

The temptation proved irresistible. When the next tip arrived, he smuggled £100,000 out of the bank and placed his bet. The next few hours seemed an eternity. When the time of the race came, he retired to a quiet corner and listened for the result on his pocket radio. The horse won. He collected his winnings, replaced the money he had 'borrowed' and handed in his resignation. He was rich. But how had the tipster managed to pick out an unbroken succession of winners?

Overcome with curiosity, he insisted on meeting the man before handing over his share. He turned out to be a small-time swindler who knew nothing about racing. He employed a string of home typists to do his clerical work. 'But how did you predict the winners?' asked the bank clerk. 'Why did you send your tips to *me*?'

The tipster told him: 'I picked 80,000 names out of the telephone directory. Yours happened to be one of them. I divided them into five groups, took a five-horse race and sent each group of 16,000 a tip for a different horse. I dropped the losers, split the winners into five more groups and sent them tips for the five horses in another race. And so on. You're the only winner in the last group of five—the last survivor of 80,000 mugs.'

In some people's lives, chance plays the part of the tipster. It gives one person a run of wins in a lottery. It enables another to retire after a couple of week's play at craps. Yet for every person favoured by chance, another must be cheated. Otherwise, it would be impossible to calculate odds and casinos would go out of business. Most punters understand this at gut level and like the doctrine of divine election, yet it brings two different responses. Some believe that nothing can influence the workings of chance and hope that they will win through a pure act of grace. Others feel that they must justify themselves by works and deserve to win only if they do everything they can to improve their chances. Hence the obsession of the puritanical British for permutations or 'perms' when filling in their football pool coupons which carry the promise of some of the biggest rewards in world betting. Every year, thousands collect five-figure payments or 'dividends' and a few win much more. Mrs Nell Fletcher of Argyllshire, who picked up a record £680,697 in February 1974, was the tenth punter to scoop a jackpot of more than £500,000.

A number of promoters offer a variety of bets with stakes as low as one penny for twenty attempts, but easily the most popular is Littlewood's 'treble chance' which pays out over £1,000,000 each week. Entrants are required to place a x against any 8 of some 54 listed matches. For each marked match ending in a draw in which goals have been scored ('score-draw'), they gain three points; for every 'no-score draw', two points; for every match postponed, declared void or won by the 'away' team, one-and-a-half points; and for every match won by the 'home' team, one point. After tax, expenses, and profits have been deducted, the pooled stakes are divided among those who have gained the most points, the maximum being twenty-four for eight score-draws.

Though some punters use form books to help them, it is impossible to predict with any degree of certainty whether a match between two equally matched teams will end in a score-draw, a no-score-draw or a narrow win for one or the other. Finding *eight* score-draws is for all practical purposes a matter of pure chance and most punters realize this by marking, say, every seventh match as it appears on the coupon or choosing according to their home street number and family birth dates. Such methods regularly pay off. One woman told how she placed her coupon on the kitchen floor, danced round it naked, threw a handful of split peas in the air and marked the matches they fell on. An Irish labourer who picked up more than a quarter of a million pounds chose matches involving teams for towns in which he had worked and several winners have admitted picking their winning list of games out of a hat.

Many, however, hope to improve their chance by using elementary number techniques. They pick out a larger number of matches which they think may end in score-draws and make a series of separate entries, or lines, covering all the possible combinations of eight matches from these. Even if most of the matches chosen do not end in score-draws, they hope that at least eight will do so, ensuring a first dividend. These 'perms' are actively encouraged by the promoters who print their own systems on entry coupons and even provide special charts for checking them when the results are known. Vernons, for instance, offer a 'Lucky V-Plan' costing £1. To make 800 separate entries at one-eighth of a penny each, the punter puts a x in the box provided and marks eighteen matches from the full list. If 8 of the first 10 marked end in score-draws, he is sure to get the maximum of 24 points, and the same applies if 5 of the first 10 and 5 of the last 8 end in score-draws. Nine scattered score-draws ensure that at least 7 will fall in 1 of the 800 lines, guaranteeing at least 22 points if the eighth is a home win, and possibly more. Any of these outcomes would qualify for a dividend.

Even so, the odds against winning are vast, for to cover all the possible combinations of 8 from a list of 55 would need 1,217,566,350 lines. Even with entries costing as little as one-twentieth of a penny, none of the promoters offer dividends anything like big enough to make a profit. Yet one punter at least sent in a coupon in the absolute certainty that it would win.

In 1970, a dentist of Middlesex, England spotted a loophole in the rules, worked out a plan of campaign and waited patiently for a chance to put it into practice. It came on a Saturday in February. As the day went on the weather got steadily worse and every hour brought news of more games being called off. By 3.45 p.m., he had compiled from his radio a list of thirty cancelled matches. He telephoned a pools agent in Liverpool where Vernons have their headquarters. The agent was standing by with a coupon and a blank cheque. At the dentist's instructions, he entered a permutation that covered all the remaining matches, filled in the cheque and handed them both in at Vernons ten minutes before the 4 p.m. deadline. The stake was £1,600 but the dentist was on 'a winner to nothing'. He collected nine first dividends and numerous smaller dividends, making a total of £47,464. His coup cannot be repeated. Vernons changed their deadline to 3 p.m. and postponed matches are no longer eliminated but are automatically scored at one-and-a-half points.

Despite their specious appeal to mathematical skill, football pools are more like lotteries than they have ever been. Other games combine chance with aptitudes that have nothing to do with number. In bridge, it is interesting to know that our chances of picking up a complete suit are 158,753,389,899 to 1 against but experience, intuition, and a memory are at least as important as a knowledge of the odds. In poker, it is interesting to know that we have only one chance in 649,740 of being dealt a royal flush, one in 72, 193 of a straight flush, one in 694 of a full house, one in 509 of a flush, one in 138 of a pair, and one in 255 of a straight, but if we hope to make a killing, we need insight and acting ability.

There is one game, however, where an understanding of number can help us not just to minimize our loses but to make a steady profit. In casinos, it will even enable us to beat the house advantage. This is blackjack, *vingt-et-un* or pontoon for which Dr Edward O. Thorp, Professor of Mathematics at the University of California at Irvine has worked out a system which swings the balance of probability decidedly in favour of the player against the bank.[16]

The rules of blackjack vary from place to place but usually, players are required to make their bets before reciving cards. The dealer then deals two cards to each and two to himself. Whether the players' cards are dealt face up or face down depends on the casino but one of the dealer's cards is always dealt face down. The aim of the game is to build a hand with a higher number of points than the dealer's but not exceeding 21 ('busting'). Court cards count 10, aces 11 or 1, the rest their face value. A total of 21 consisting of an ace and a 10 or court card is known as a blackjack, 'natural' or pontoon.

After seeing his cards, each player in turn may ask the dealer for a further card or cards by saying 'hit' or 'twist'. When he thinks he has come so close to 21 that a further card will make him bust, he says 'stand' or 'stick'. If he does bust, he reveals his cards, throws them in and loses his stake. When all the players have stood or bust, the dealer reveals his second card and if the two total 16 or less, he must, by the rules of the house, deal himself another and keep on doing so until he reaches a total of 17 or more. He must then stand. Players still in the game reveal their cards and those with a total lower than the dealer's lose their stakes. Those with a greater total win a sum equal to their stakes, except for those with naturals who collect one-and-a-half times as much. If the dealer has a natural, this beats all other combinations. When a player and the dealer each have a similar total or a natural, they tie and the player keeps his stake but does not win anything in addition.

Two other bets are possible. After seeing his cards, a player may turn them up, double his stake and call for one, and not more than one, further card which remains face down. This is called 'doubling down'. Secondly, if the turned-up card of the dealer is an ace, any player may make a side-bet of an amount equal to half his original stake. The dealer then turns up his other card and if it is a ten or court card, giving him a natural, the player wins a sum equal to twice that of his side-bet. If not, he loses the side-bet. This ploy is known as 'insurance', since a turned-up ace gives the dealer a fair chance of a natural and when a player knows that he himself does not have a natural and is therefore likely to lose to the dealer, he can recoup his likely loss by the profit on his side-bet. After an insurance bet, play continues in the usual way.

Success in blackjack clearly depends on choosing the right point at which to stand and to do this, a player must take into account both the cards in his own hand and the turned-up card of the dealer. Generally speaking, the higher this is, the better the dealer's chances of getting a high total without busting and the higher the total at which the player should aim before standing.

On this assumption, Dr Thorp developed a 'basic strategy' with the help of mathematical friends and the use of an IBM 704 computer which the Massachusetts Institute of Technology Computation Centre made available. He worked out the decisions most likely to prove favourable for the player, giving him, claims Dr Thorp, an advantage that can be as high as one per cent and even with the most unfavourable house rules, reducing the casino advantage to one per cent.

He did not stop at the basic strategy, for the very nature of blackjack places it in a different category from craps or roulette. In these games, the odds against any specified number are precisely the same at each throw or turn of the wheel, irrespective of previous outcomes, because dice and roulette marbles do not have memories. Cards do, at least in blackjack when the dealer plays straight through the pack without shuffling. After four aces have been used up, the chances of another appearing are zero and the odds against other cards turning up vary according to the number of similar cards that have already been played. If a punter keeps track of past cards, assesses the balance of probabilities remaining and adjusts his bets accordingly, he can swing the advantage still more in his favour. Dr Thorp worked out several systems for doing this by combining simple methods of counting key cards with strategies based on computer calculations of the remaining probabilities.[17]

As we have seen, gambling systems tend to be as reliable as pyramid selling schemes, but Dr Thorp gives us good reasons for thinking that his might be an exception. He first tested it while teaching at the Massachusetts Institute of Technology after a *Boston Globe* reporter read the abstract of a paper about it which he planned to read at the annual meeting of the American Mathematical Society in Washington, D.C. An interview he gave was widely syndicated and among the hundreds of enquiries that poured in was a proposal from two millionaires who offered

$10,000 backing for a trial run. Next spring, the three of them took a trip to Reno, Nevada and in thirty hours of play, turned the $10,000 into $21,000 despite cheating dealers and other obstacles placed in their way by worried casino owners.[18]

After the first edition of Dr Thorp's book was published, Las Vegas casinos changed their rules because they were losing too much money to punters playing his system. Soon, however, they were changed back for there was no effective defence against some of his strategies. Even so, consistent winners have to expect counter-measures which may include cheating, constant shuffling of the pack, the use of two or four packs instead of one and even a refusal to let them play. None of these need worry the really keen player, however, for he can fairly easily disguise the fact that he is using a system and if necessary, disguise himself.[19]

It may be asked how the casinos find it worth their while to continue with blackjack, if anyone can learn to win by simply reading a book. But do punters necessarily wish to win? It seems likely that many gamble for the social cachet which they think it brings, for excitement that thrives on *un*certainty and perhaps to act out fantasies ranging from the magnanimous to the masochistic. After one disastrous loss, the novelist Dostoevski had a spontaneous orgasm. Probably most gamble for fun. They know that an appreciation of number can improve their chances but they do not take their gambling all that seriously. They simply cannot be bothered with a system like Dr Thorp's which needs study, practice, and nerve.

6

Lies, Damned Lies, and Statistics

LIKE gamblers, we all try to manipulate numbers to our advantage in the everyday affairs of life. Business, politics, and medicine are at least partly number games and so are weather forecasting, archaeology, and hotel management.

When we take out insurance, we are really placing a bet. The company acts as bookmaker, our premium is the stake and we are gambling on whatever risk we are insuring against. With 'term' insurance, the arrangement by which our heirs pick up a lump sum when we die, we are gambling on our own lives. The odds are always against us. By studying past records, actuaries predict the likely number of years we still have left, given our present age, sex, health, occupation, marital status, and drinking and smoking habits. Having worked out our form, they fit us into the company's book at odds which will cover not only the possibility of a claim but also a share of overheads and a percentage for profit.

Of course, insurers cannot be sure that we shall perform exactly as the actuaries have predicted, any more than a bookmaker can be sure of a horse's performance on the race-course. Occasionally, so many favourites run backwards that claims exceed premiums and the company welshes. Mostly, though, it manages to pay out on winners, cover overheads and make a profit from its customers as a whole. For us punters, insurance is not the best of bets. Our only hope of beating the book is to drop dead the day after signing.

In other life gambles, too, the rules are similar to those that apply on the race-course or in the casino. Probability is again decided by the number of favourable cases divided by the number of possible cases and this applies particularly in the lottery of birth. Since babies can be either boys or girls, would-be parents have a one-in-two chance of getting a boy and an identical

99

chance of getting a girl.* The 'maturity-of-chances' fallacy is still a fallacy. A couple who have already had five boys may think it a fair bet that the sixth will be a girl, forgetting that each birth is a separate event unconnected with the rest. Their chances of coming up lucky, whatever they conceive their luck to be, are still what they always were, one in two.

The sex of children is the subject of another classic fallacy. If a father says, 'At least one of my two children is a boy,' and goes on to ask, 'What are the chances of the other also being a boy?', we instinctively answer 'one in two' which is wrong. We are assuming that the *first* of his children is a boy and that the only possibilities to be considered are boy–boy and boy–girl. In fact, there are three possibilities, boy–boy, boy–girl, and girl–boy, so the chances of the other child being a boy are not one in two but one in three.[1]

Surprisingly, though, a young man who tells his girl, 'You're one in a billion' is not necessarily flattering her. He could well be a gambler with a shrewd appreciation of the odds. As the statistician M. J. Moroney pointed out, she could have:

a Grecian nose—probability 0·01;
platinum blonde hair—probability 0·01;
each eye a different colour—probability 0·001, and
an excellent knowledge of statistics—probability 0·00001.

Multiplying these together, we get a combined probability of 0·000 000 000 001, which is one in a billion (defined as a million million). Whether he was on to a winner depends on what you think of Grecian-nosed, platinum blondes, with eyes of different colours, and minds like calculating machines.[2]

Few people appreciate how probability works in practice and if we announce at a party with sixty guests that at least two of them share the same birthday, i.e. were born on the same date of the same month, they will probably be sceptical. Yet a check will almost certainly show that we are right.

To see why, let us (a) forget about Leap Year and (b) assume that births are evenly distributed throughout the year. If we take only two people, the chances of the second having a *different*

*Statistically this is not strictly accurate. In Britain, for instance, some 106 boys have been born for every 100 girls in recent years.

birthday from the first are clearly 364 out of 365 or $\frac{364}{365}$. Since two dates have now been accounted for, the chances of a third person having a different birthday from either are $\frac{363}{365}$ and by the time we come to the twenty-fourth person, they are down to $\frac{342}{365}$. To work out the chances of all twenty-four having different birthdays, we multiply the fractions together, and by reducing get a result of $\frac{23}{50}$. The chances of them *not* all having different birthdays are therefore $\frac{27}{50}$ or better than evens. With thirty people, the chances of at least two sharing a birthday are better than 2 to 1 and although we should need 366 people, one more than all the possible birthdays available, to be absolutely sure that two coincided, the chances come very close to certainty with as few as sixty.[3]

If this seems unlikely, we can make a quick check by writing down at random a list of well-known people and looking up their birth (or death) days in a reference book. In the succession of American presidents, we have only to go as far as the twenty-ninth to find a coincidence of birthdays. Warren G. Harding shared 2 November with James K. Polk, the eleventh president. As for deaths, there was a coincidence of three among the first five presidents. John Adams, the second, died like Thomas Jefferson, the third, and James Monroe, the fifth, on 4 July.[4]

Since the Norman Conquest in 1066, England has had 43 rulers. Not all their birthdays are recorded but we find King Edward I sharing 25 April with Oliver Cromwell, Lord Protector of the Commonwealth.

We have checked the working of our birthday example by drawing on statistics, which are simply numerical facts that have been collected and arranged in a particular way. No civilization could exist without them. Even the most primitive tribe needs to know how many mouths have to be fed and how many hunters are available to feed them. In the Bible, we read that both Moses and David conducted censuses and another took place just before the birth of Jesus. The Babylonians kept elaborate records of the movements of the heavenly bodies, the Egyptians of crops, taxes, and the flooding of the Nile. Nowadays we rely on statistics not just in government, technology, and science but in warfare, advertising, and even the arts.

Let us look at some of the ways in which we use them:

During the Second World War, statistics saved the lives of thousands of British and American seamen and millions of tons of shipping. A detailed study of submarine attacks on North Atlantic convoys revealed an interesting link between the number of ships sunk (r), the number of ships in the convoy (N) and the number of escort vessels (C). If k was the constant that related them, it always happened that $r = \dfrac{k}{NC}$. It seemed, then, that the proportion of losses could be cut either by increasing the number of ships in a convoy or the number of escort vessels or both. This proved to be no idle theory. In practice, one big convoy was indeed found to suffer proportionately fewer losses than two smaller ones. Hence the saving in lives and ships.[5]

Marketing men often need to know the likely effectiveness of different strategies when planning a promotion. Again, statistics help. A food manufacturer who wishes to launch a new product or boost sales of an old one may decide to circulate discount coupons ('5p Off') which entitle customers to a lower price if they are handed over when the new product is bought. But how many will be redeemed? Does it make any difference how they are distributed? Figures compiled over many years by firms selling mass-market products show that between 20 and 40 per cent of the coupons will be redeemed if they are delivered to individual homes. The redemption rate of identical coupons printed in newspapers from which they have to be cut is only 1 per cent.[6]

As for the arts, the critic L. A. Sherman told a disbelieving world as long ago as 1892 that '23·43 was the resultant of the forces which had made Macaulay's literary character'. By this he meant that he had worked out the average length of Macaulay's sentences and had found that it came to 23·43 words. Since then, the statistical analysis of language has mushroomed, especially with the advent of the computer. It is now known that the English language has some 340,000 different word-types (win, wins, won, and winner are separate types), though our longest word-list, the *Oxford English Dictionary*, contains only 300,000 and no writer is likely to use more than a tenth of these in his life's work. Even Shakespeare used only 29,066 different words and these include 90 per cent of those found in any work of literature, even today. In any passage of English, the definite article is likely to account for 7 words in every 100 and with the next nine commonest words, which vary from genre to genre, for a further 24 in every 100. From studies of this kind, it is possible to compile a norm for any genre or period and hence to show how individual authors deviate from it.[7]

Dr Yehuda Radday of the Israel Institute of Technology applied this technique to the Book of Isaiah. On literary grounds, it has long been attributed to at least two separate authors, 'first Isaiah' who dates from the end of the eighth century B.C. and a separate hand, or hands, who wrote the rest. Dr Radday's aim was to see whether the differences between chapters 1 to 39 and the remainder of the book could be quantified. He divided the first 39 chapters into three sections and the rest also into three sections. He worked out the lengths of the sentences in all six separately and noted the frequency of different parts of speech and various peculiarities of style. He then ran his results through a computer. The print-out showed clearly that the chances of chapters 1 to 39 having been written by the same author as the rest were 100,000 to 1 against.

We could multiply examples indefinitely but perhaps it would be more useful to take a look at some of the ways in which statistics are *mis*used. Most of us already have a healthy prejudice against them, know only too well that there are 'lies, damned lies, and statistics' and are quite sure that if all the statisticians in the world were laid end-to-end across the bed of the Pacific Ocean, it would be a very good thing indeed. Yet we need to discriminate. Though much of our distrust is justified, it is not always fair to blame statisticians. The real culprits are usually the ignorant, the lazy, and the unscrupulous who misapply the figures with which they have been supplied. Often, we have only ourselves to blame for being so gullible.

Perhaps our commonest fault is to accept uncritically any figures quoted by someone claiming to be an 'expert'. On reflection it is obvious that statements like 'America has 500,000 schizophrenics' or '400,000 Britons are alcoholics' are sheer nonsense because there is no generally agreed definition of either condition and to put a figure on anything so nebulous is like publishing a census of fairies.

Even figures that *could* be true and come from seemingly impeccable sources may be just as worthless. One research organization sponsored by the British government publishes details of the technical assistance it gives to industry. They look impressive in the annual report but how are they compiled? A friend of mine who works as a consultant for the organization told me, 'At the end of each week, we have to account for every hour of every day so that it can be charged to a client's account. Now suppose I go out for a haircut, meet some friends and spend a couple of hours drinking with them. I can't put down "Monday—2–5 p.m.—haircut and drinks". I write instead, "Research on Project X for Y company". The bill for those three hours goes to Y company who charge them to their Project X account. And that's how they appear in the annual statistics.'

If this sounds frivolous, it is worth considering the British government report entitled *Accuracy of Certification of Cause of Death*. This compares the results of post mortems on 9,501 patients in seventy-five hospitals during 1959 with the clinical diagnoses of the underlying causes of death. Post-mortem findings and clinical diagnoses agreed in only 45·3 per cent of the cases. The clinicians tended to diagnose lung cancer and peptic ulcer less frequently than the pathologists, though they usually agreed with post-mortem findings on pneumonia, bronchitis, and arteriosclerotic heart disease. The clinicians also tended to 'overdiagnose cerebrovascular disease'. The report commented: 'Only one quarter of deaths, however, were associated with disagreements of fact and these tended to occur more frequently when the clinician was less certain of his diagnosis, or in deaths of older patients.'[8]

These findings need cast no reflection on the doctors involved, for old people especially may suffer simultaneously from several potentially fatal illnesses and it is a matter of luck which kills them first. A busy houseman or G.P. may think that his time is better spent in tending the living than fussing over the exact cause of death, especially when he is unable to diagnose it with any degree of accuracy. Yet these findings make nonsense of official cause-of-death statistics. From the scientific point of view, it is hard to imagine a more fatuous statement than '*only* one quarter of deaths were associated with disagreements of fact' (my italics). All the statistics tell us is what doctors write on death certificates and this may or may not be the truth.

One wonders how many other figures are equally worthless for in the last analysis, each of the billions of facts that are collected each year depends for its reliability on an individual human being who, being human, may have made a mistake. In the computer business, there is a catchword GIGO, meaning 'Garbage In, Garbage Out'. In other words, if you feed nonsense into a computer, you can only expect nonsense to come out of it. The same applies to statistics generally, as we should do well to remember.

Even if the figures are meaningful and accurate, they may lead us astray because they are incomplete or taken out of context. In 1946, there were 4,712 deaths on American railways. Used by airlines or bus operators in comparative advertising, such a figure could be used to imply that anyone boarding a train was buying into a game of Russian roulette. But what are the facts behind the figures?

> 'We find that of the 4,712 only 132 were passengers. There were 736 railroad employees, and 1,618 trespassers, mostly riding the rails. In addition, 2,025 people were killed at grade crossings. Under the heading "other non-trespassers" there are listed 201 fatalities. This analysis of the figures (given out by the National Safety Council) is clear provided that we understand the meaning of the terms used. For example, passenger fatalities include not only those travelling on trains, and those getting on or off trains, but "other persons lawfully on railway premises".'[9]

Manipulators do not need to falsify or curtail statistics to make them misleading. They can present accurate figures in a way that gives a highly misleading impression. A common way of doing this is by use of what Darrel Huff has called 'the gee-whizz graph',[10] which makes an insignificant, possibly temporary movement alarming, usually by monkeying with the vertical scale. As an example, let us take two graphs based on the *Financial Times* Ordinary Share Index for March 1976. Figure 1[11] shows a plunge in share prices that could easily stampede nervous investors into jamming their brokers' lines with calls to 'sell the lot'. Figure 2[12] gives the same information set in a wider context and suggests, rightly, that the March fall was a small fluctuation in a trading range that varied between 390 and 420. In this particular case, the gee-whizz Figure 1 was not intended to mislead but it shows how careful we have to be. The higher the starting point of the vertical scale, the easier it is to be misled.

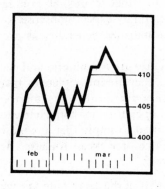

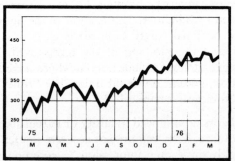

Figures 1 and 2 The same information in different contexts

Ignorance is another bugbear. We cannot expect to understand the significance of a set of statistics unless we have enough background knowledge to ask the right questions. Shareholders may be delighted to hear from a company chairman that sales have increased by 20 per cent. But compared with what period? Last month, last quarter, last year or the average of the years 1930 to 1940? Does the 20 per cent refer to value? If so, it may hide a sharp decline in volume in a period of inflation. Even if volume has increased, it could be because future sales have been pre-empted by a whizz-kid sales director who persuaded customers to overstock before moving on to a higher-paid job elsewhere on the strength of his apparent success.

Even a genuine increase in volume may be disastrous if it was achieved by lowering prices to an unprofitable level. With news-papers and magazines, a rising circulation is not necessarily the high road to increasing profits. Advertising normally brings in more money than circulation, so that each copy is heavily sub-sidized. At least one national magazine had to close down because the extra readers attracted by the flair of its editor were not the suggestible, free-spending types favoured by mass-market advertisers. Page rates could not be put up and there came a point where every extra copy sold brought not a profit but a loss.

Extrapolation is the stickiest trap of all. It means working out from known facts a series of deductions on the assumption that the pattern will continue. Often these deductions are themselves treated as facts but since they deal with the future, they can be nothing of the kind. To make an extreme example, we know that the sun has risen every morning since man started taking observa-tions and no bookmaker would offer even the longest odds on its failing to rise tomorrow. In practice, he would probably be right, but we can never be *sure* of the sun's future appearances if only because it might be destroyed overnight in some cosmic disaster which we have no means of foreseeing.

Planners working on projects that will come to fruition many years ahead work by extrapolation and often make huge errors as a result. This is inevitable because of the unknown quantities which assume ever greater proportions as civilization becomes more complicated. Who could have foretold the rise in oil prices engineered by O.P.E.C., the failure of the Russian wheat crop in 1974 or the sudden awareness of the British trade union movement of its power to make or break the economy? Yet all these events have affected every single person in the world.

What were once thought to be rational decisions based on a cool appraisal of the statistics turn out to be wild gambles. British Post Office statisticians predicted that there would be a boom on long-distance telephone calls and prudently a new telecommunications building costing £650,000 was planned for Rochdale, Lancashire. By the time it was completed in mid-1976, huge increases in call charges had cut demand to a point where the new exchange was no longer needed and it was put up for sale as office accommodation.[13]

Similarly, Britain's nationalized Central Electricity Generating Board forecast in 1964 that demand for electricity would grow by 4 per cent a year and built the power stations needed to meet it. In the event, growth averaged only 2½ per cent a year, and by 1976 the board found that it could produce 40 per cent more power than it could sell. Ironically, the situation was made worse by a 'Save It' campaign run by the government to persuade the public to cut down on its overall use of power when the rise in world oil prices adversely affected the balance of payments. The C.E.G.B. was forced to close down twenty-eight power stations, some of them only twenty years old.[14]

A pessimist might conclude that society is now so complex that it is impossible to make reliable extrapolations from statistics because there are so many imponderables and the world has become so interdependent that an unforeseen happening in a seemingly remote country can make nonsense of the best-laid plans. Yet the collection of statistics has become an obsession in every advanced country. They pour in on us in an ever-rising flood which, like the sorcerer's apprentice, we seem unable to stem. Aided by computers capable of millions of calculations per second, we examine them as expectantly as soothsayers of old peered at the entrails in hopes of finding guidance for the future.

Students of cycles are among the most diligent of our latter-day seers. According to Dr Edward R. Dewey, president of the Foundation for the Study of Cycles which is affiliated to the University of Pittsburgh, there are thirty-seven separate stock-market cycles or hints of cycles. Perhaps the most important is a 9·2-year cycle in prices which, he said in 1971, 'has repeated fourteen times since 1834, two years before the siege of the Alamo and ten years before Samuel Morse sent his first message over the telegraph line.' The odds against this happening by chance are some 5,000 to 1 according to a standard probability test.[15]

The cycle peaked in 1965 and was due to peak again in 1974, the year of the greatest crash since 1929. Yet this, says the cyclists, does not invalidate their theory. Cycles are only one of several factors affecting stock prices, others being long-term economic trends and the effects of inflation. It does mean, though, that investors should never put their shirts on the 9·2-year cycle, not to mention those of 18·2 years, 6·01 years, 41 months, and 17·16 weeks. If they do, they may be letting themselves in for a long cold winter.

Cycles have also been traced in commodity prices. One of the first chartists to study them was Samuel Turner Benner, a Bainbridge, Ohio farmer who was ruined by an outbreak of hog cholera that coincided with the 1873 business panic. He turned his mind to higher things and two years after going bankrupt, published *Benner's Prophecies of Future Ups and Downs in Prices* which he repeatedly updated until 1907. His discovery of a pig iron cycle with periods of eight, nine, and ten years for crests followed by nine, seven, and eleven years for troughs held good until the Second World War. 'Had you traded from 1875 to 1935 on the basis of Benner's cycle you would have made forty-four times as much as you lost', said Dr Dewey.[16]

He himself studied copper prices between 1824 and 1973, and discovered that they rose and fell in waves lasting precisely 9·15 years. These came so regularly that a family who bought and sold accordingly would have made a killing in fourteen cycles out of the sixteen studied. The total gain, assuming a fixed investment in each cycle and no brokerage, would have been 754 per cent.[17]

Other commodity cycles are said to range from 3·5 to 3·75 years for corn to 17·75 years for cotton, from 26·64 months for oats to 16·67 years for English wrought iron. This last cycle persisted from 1288–1908, almost as long as the 54-year cycle in European wheat prices discovered by Lord Beveridge in a study covering the years between 1500 and 1869. It has since been traced back as far as 1260 and forward to 1940. Dr Dewey points out, however:

'. . . This does not mean that the crests and troughs come exactly 54 years from each other. The actual highs and lows are distorted, one way or another, by randoms and other cycles. There is, however, a *tendency* for *areas* of strength to follow each other at 50 to 60 year intervals and for such areas of strength to be separated by corresponding areas of weakness.'[18]

The note of caution is refreshing. Some cyclists (or should we call them trick-cyclists?) give the impression that the world's economy works like a watch, ticking off slumps and booms with unfailing regularity. It is hard to resist the impulse to ask them: 'If you're so smart, why aren't you rich?'

On the other hand, it would be quite wrong to dismiss the idea of cycles completely. More than half a century ago, the Russian economist Nicolai Kondratieff pointed out that periods of rising food and raw material prices had set in at intervals of around fifty years since the eighteenth century. Upswings in the Kondratieff cycle seem to have started at the end of the eighteenth century, halfway through the nineteenth, at the end of the nineteenth and, a little early perhaps, in the late 1930s. Prices started to fall in 1815, 1873, 1920, and 1951.

W. W. Rostow, Professor of Economics and History at the University of Texas at Austin, has linked the upswings with population rises in Europe in the early 1790s and in the mid-1840s, and in both America and Europe in the mid-1890s and the mid-1930s. Similarly, when the wheel came round full circle in the mid-1970s, we could point to the pressures of third-world nations and their insistence on securing for themselves a greater share of global resources by raising oil prices.

Professor Rostow wrote:

'I am not wedded to the notion that these cycles will continue in the future. But I would guess that the inexorable pressure of excessive population increase in the developing world, the tendency of the poor to spend increases in income disproportionately on food, the rising demand among the more affluent for grain-expensive proteins, the pace of industrialization among those catching up, and the strains of the energy crisis will persist.

'Given these powerful and sustained demands operating on food, energy, and raw material prices and the costs we shall have to incur to achieve and maintain clean air and water, I believe we are in for a long period when the prices of these basic inputs to the economy will remain relatively high.'[19]

Weather patterns may be even more crucial to world prices. As we saw in chapter 2, those associated with the 11-year sunspot cycle can have far-reaching effects and in late 1974, British scientists at the government-financed Appleton Laboratory near Slough, Buckinghamshire reported a clear relationship with world wheat harvests and with yield per acre of potatoes, swedes, and turnips in Britain. These tended to be higher in the years of sunspot maxima.[20]

In 1974, the sunspot cycle was hovering around its minimum and the Appleton findings were given further confirmation the very next year which for Britain turned out to be the fifth driest of the century. Potato yields were down by a third and in early 1976, prices in the shops more than doubled, entirely because of lack of rain during the growing season.

The link with sunspot cycles is only one of many patterns that have been found in weather statistics, some of them trivial, others of great importance. To take the trivial first, it has been found by the Australian meteorologist Dr E. G. Bowen and confirmed by the U.S. Weather Bureau that there is a tendency for heavy rain to fall in particular places on particular dates. The citizens of Brisbane know that they will almost certainly be deluged each 23 January. No one has yet found the reason.[21] In Britain, amateur meteorologist George Nicholson checked London's weather records for the fifteen years between 1954 and 1969 and found that one day of the week regularly had more rain than the others. The rainy day was Thursday with downpours on 370 Thursdays out of the 780 investigated. Mr John Harding, head of the hydro-meteorological section at the British Meteorological Office at Bracknell, Berkshire confirmed Mr Nicholson's findings but was unable to explain them. 'It could be air pollution,' he said, 'but there seems no logical explanation at all.'[22]

Nor do we know why cold wet Junes tend to follow mild Februaries and vice versa. Dr F. H. W. Green of Oxford University discovered this freak of European weather when he checked records going back at least thirty-five years in Norway, Denmark,

Scotland, and England. Moreover, he found that the area affected
was gradually spreading to the south and west. It has long been
known that westerly cyclonic winds bring mild winters and cool
summers, whereas easterlies waft cold air over north-west Europe
in winter and warm air from the continental land mass in summer.
But why June wind directions should in any way be linked with
those of February is far from clear. Dr Green has suggested that
it may be connected with the southward spread of the Arctic high-
pressure region in recent years, but once again we do not know
why this is spreading.[23]

Another mystery is why weather patterns should be linked
with odd- and even-numbered years. Yet such seems to be the
case in Britain, at least during the period 1880 to 1967. Mr N. E.
Davis of the Meteorological Office, Strike Command, High
Wycombe, Buckinghamshire examined weather records at more
than a score of places from the Outer Hebrides to the Isles of
Scilly. He found that warm, sunny summers ran in two inter-
weaving patterns. In odd-numbered years, they rose to a maximum
every ten or a dozen years, with excellent summers occurring in
1947 and 1959. In even-numbered years, they reached a maximum
every 34 years, so that there were runs of better-than-average
summers in 1898–1906 and 1932–40, and abysmal summers in
1888, 1922, and 1956. It must be said that British summers since
1967 have not followed this pattern very closely. This could be
a statistical aberration. Alternatively, it might be thought that if
we look long enough at any bunch of statistics we shall eventually
find some pattern in them.[24]

The effects of some of these unexplained patterns may be
considerable. Variations in rainfall lead to fluctuations in the
water levels of the Great Lakes of America, and Colonel Harry A.
Musham of Chicago claimed to have discovered a cycle of 22·75
years in the levels of Lakes Michigan and Huron. In 1941, he
foretold that Michigan would reach a 'high' in 1951 and 1952 and
two years later predicted a 'low' for 1964. Both predictions proved
accurate. In 1964, the Lake levels reached the lowest point ever
recorded and the economic effects were so alarming that Congress
held a formal investigation. A fall of one foot restricted some
United States lake ports so severely that shippers reckoned their
losses at $7 million. Power supplies were also affected. Because

of reduced outflow from the Lakes, output of the state generating plant at Niagara alone produced 1¼ billion (U.S.) kilowatt-hours less than average in the first six months of the year.[25]

More important than any of these smaller cycles, however, are those of the so-called ice ages which have cataclysmic effects on the whole world. Few, if any, meteorologists would deny that such a cycle exists but they disagree violently about the length and spacing of the phases and also about their effects.

One of the longest of these ice-age cycles was suggested by Professor W. H. McCrae of Sussex University, England who linked it with the swing of earth and the rest of the solar system through dense clouds of gas or 'dust lanes' fringing our galaxy. Analysis of soil samples taken from the moon suggested that this might happen every 100 million years, the approximate interval between our last two major ice ages. The second of these lasted at least 2,500,000 years and ended a mere 10,000 years ago. On this evidence, there is no immediate hurry to invest in a fur topcoat.[26]

Within each major ice age, however, there were wide fluctuations of temperature. In the warmer spells, mean temperatures were sometimes slightly higher than those of today, though they dropped to as much as 10 degrees centigrade lower during the so-called glacial periods, enough to bring the Arctic ice-sheet as far south as Cincinatti, Ohio and St Louis, Missouri, while smaller glaciers extended to Southern California, London, England, Amsterdam, Holland, and Moscow in the U.S.S.R. There were also wide fluctuations in temperature in the periods between the ice ages.[27]

How our present situation relates to the past is even more uncertain. Some experts point to the record of the glaciers which seems to show that earth's temperature peaks approximately every 100,000 years and since the last maximum was reached some 5,000 years ago, we have already started on a downward trend which can be expected to bottom out some 50,000 years from now.[28]

Narrowing the range still further, Professor Willi Dansgaard of the University of Copenhagen in Denmark has looked at the prospects for the next century or so. It is known that the proportion of heavy oxygen atoms in water increases with temperature and by analysing the layers of ice laid down annually in the polar regions, it is possible to tell whether each year was relatively warmer or colder. After studying a 2,400-foot core covering 150,000 years which American Army engineers drilled from the Greenland ice, Professor Dansgaard believed that he had found two intermeshing cycles of 80 and 180 years. He concluded that our prospects are chilly until the mid-1990s but after that, those of us who are still alive can look forward to a balmy old age before the onset of another cold period due to a turndown of the 180-year cycle in the twenty-first century.[29]

Broadly speaking, Professor Dansgaard's findings confirm those of the distinguished British meteorologist Hubert H. Lamb. After studying London weather records back to 1669, ships' logs, diaries, state and ecclesiastical archives, he was able to form a good idea of British weather back to A.D. 1100. He found that a cold period got under way towards the middle of every second century, for instance in the late 1700s, in the late 1500s (when the Thames froze over) and in the late 1300s. Obviously some years or even runs of years will be exceptions but on the whole, he believes that the pattern is repeating itself in the second half of the twentieth century.[30]

Other researchers have pointed out that in the last seventy-five years, the world has enjoyed weather that has been warmer on average than that of 95 per cent of the last 700,000 years and a cooler period probably lies ahead. The possibility has even attracted the interest of America's Central Intelligence Agency which stated in a report released in May 1976, 'it is increasingly evident that the Intelligence Community must understand the magnitude of international threats which occur as a function of climatic change.'[31]

The C.I.A. bases its concern largely on the work of Professor Reid Bryson and his colleagues at the Institute of Environmental Studies at Wisconsin. Assuming that a fall in temperature is likely in the northern hemisphere, they point out that a difference of only one degree centigrade would push far to the south the westerly winds that bring so much rain to Britain and north-west Europe. The climate of northern latitudes would become colder and drier, making it impossible to grow food in what are now highly productive areas of North America, northern Europe and the U.S.S.R. In Europe, the colder climate alone would cut the number of people that could be supported by each hectare of arable land from three to two. In China, only four people could be supported per hectare, against the present figure of seven. In addition, the monsoon would no longer bring rain to vast areas of North Africa, northern India, and southern China which currently benefit from them.[32]

The military significance of this is obvious. The rise and fall of some of the great civilizations of the past can be linked with climatic changes and there is every reason to think that the pattern will continue in the future. Starving nations could well be forced to embark on wars of aggression simply in order to survive and it is unlikely that they would be deterred by a nuclear balance of terror. Hence the interest of the C.I.A.

Happily, though, the outlook may not be so alarming. For one thing, not all meteorologists agree that we are heading for a 'little ice age'. They believe that their colleagues have made unwarranted inferences from the statistics which in any case are far from precise. In March 1976, Dr B. J. Mason, director general of the British Meteorological Office, told a meeting of the Royal Meteorological Society:

> 'There is no real basis for the alarmist predictions of an imminent ice age, which have been largely based on extrapolation of the thirty-year trend of falling temperatures in the northern hemisphere between 1940 and 1965. Apart from the strong dubiety of making a forecast from such a short-period trend, there is now evidence that the trend has been arrested . . .
>
> 'Since climates are not stationary, a particular period of say 50 years may not be representative or provide good guidance as to what may be expected in the next fifty years. In fact, the period 1931 to 1960, on the statistics of which many recent investigations have been based, was probably one of the most abnormal 30-year periods in the last thousand years.'[33]

Moreover, Professor Bryson himself has suggested that we may be able to avoid the worst effects of a climatological disaster by taking preventive measures. The cost, though, would be enormous, an estimated 75 billion (U.S.) dollars a year at 1974 prices, roughly equal to a fifth of the total annual investment of the entire non-communist world.[34]

All experts agree that we need to do much more research before we can form a reasonably sure view of what is likely to happen to the world's future climate. Comparatively little systematic work has yet been done and reliable records are often available for only a few years, far too short a time when the cycles being investigated may have a much longer periodicity. Often, meteorologists have to relate early anecdotal evidence to the present-day flood of highly detailed information that only a computer can digest and then match the result with tree rings, moon dust, or ice cores on whose interpretation there is still a good deal of argument.

This is obviously no basis on which to ask the world's governments to invest $75 billion a year. Yet too often, other figures used to justify the expenditure of huge sums are no better than those of the meteorologists. Can we *ever* discover new truths from statistics?

In one situation at least a conclusion is inescapable. This is when it follows inevitably because of the form of the proposition. For instance, if A is less than B, and B is less than C, it must be true that A is less than C. In this way, the logicians M. Cohen and C. Nagel proved beyond a shadow of doubt that at least two New Yorkers had scalps with exactly the same number of hairs.

We know that no human skull could possibly have a hair-bearing area of anything like 1,000 square centimetres nor could any single square centimetre carry as many as 5,000 hairs. From this it follows that every human being must have less than 5,000,000 hairs, and the possible number of separate hair-totals must also be less than 5,000,000.

Now the population of New York City is 7,895,563 (1970), which is greater than 5,000,000. Going back to our proposition, we find that the largest possible number of hair-totals (A) is less than 5,000,000 (B) and 5,000,000 (B) is less than the population of New York City (C). So the number of hair-totals (A) must also be less than the population of New York City (C), which means that *at least* two New Yorkers must share *one* of these totals.[35]

Another way in which statistics can be used predictively is much less foolproof but provided we know how to go about it, we can usually arrive at results with an acceptable degree of accuracy. This technique is sampling, a method of discovering information about a large group (parent group) by investigating a small group (sample) and applying the findings to the large group as a whole. Sampling is regularly used in market research, industrial quality control, and business generally.

As an example, let us take American airlines which in 1950 were worried by the steeply rising cost of inter-company accounting. When travellers booked a ticket from A to B, they normally bought it from the company whose plane they boarded at A. However, they might wish to break their journey at C, D, and E *en route*, using airlines other than the one from which they bought their ticket. These airlines then had to recoup from the seller of the original ticket their proportion of the total fare. Millions of flights in which passengers used their stopover privileges saddled all airlines with a gigantic accounting operation costing an average of $120,000 a year which was sure to keep on rising. Could they achieve a reasonably fair settlement by working out the mutual debts and credits incurred in a small proportion of inter-line flights and applying these findings to the inter-line business as a whole? In a trial period of four months, they picked out twelve per cent of the inter-line accounts and came up with results accurate to within $700 in $1 million, ample justification for a permanent switch to sampling as a basis for settlement. This enabled all airlines to save huge sums in unnecessary accounting.[36]

In modern society, sampling is obviously a useful, even indispensable technique, but it has unexpected pitfalls, as the editors of the American magazine *Literary Digest* discovered in a painful episode that took place in 1936. Hoping to stimulate circulation by correctly forecasting the result of the forthcoming presidential election, they wrote to all their readers and to all American telephone subscribers and asked them how they intended to vote. Ten million letters went out and it seemed a fair assumption that the 2,200,000 who replied would reflect with reasonable accuracy the views of the electorate as a whole. Well before polling day, *Literary Digest* had a scoop on its hands. Alfred M. Landon, the Republican candidate, would be the next president of the United States. And so he might have been, had he not been beaten by the Democrat Franklin D. Roosevelt.

Landon lost the election by more than 11 million votes in one of the biggest landslides America has ever known.

Where had the *Literary Digest* gone wrong?

There is, of course, a fundamental problem with all public opinion polls, and that is the sheer volatility of public opinion. Before an election, you can ask people how they *intend* to vote but when the time comes for them to put their X on the ballot paper, they may have changed their mind. A massive swing from Dewey to Truman during the course of the 1948 election campaign is thought to be the main reason why all the major polls wrongly predicted a Dewey victory. At the time the sample was taken, many who later voted for Truman genuinely believed that they would vote for Dewey.

There is evidence too that polls sometimes have a 'band wagon' effect, swinging 'don't knows' behind the predicted winner. In recent years, attempts have been made to spot these last-minute changes by taking repeat polls right up to election day and though results have been more accurate, it is hard to see how they can ever fully account for late switchers of whom there may be considerable numbers, certainly enough to swing an election.

The *Literary Digest*'s error, though, was more elementary. If the opinions of a sample are to have a meaningful relation to those of the parent group, the sample must be carefully chosen. The *Literary Digest*'s was grossly biased. Since the people questioned were either regular readers or owned telephones, it is pretty obvious that they were wealthier than average, excluding alike the big batallions of the unemployed and the dustbowl-fleeing Okies of Steinbeck's *Grapes of Wrath*. An element of self-selection was added in that they had the choice of replying or not replying, unlike the victims of a foot-in-the-door professional interviewer. If the *Literary Digest* team had cared to look, they would have found clear evidence of bias in the answers to one of their own supplementary questions. Subjects were asked how they had voted in the previous election when there had been 45 per cent more Democratic than Republican voters. In the *Literary Digest* sample, there had been 6 per cent more Republicans than Democrats.[37] All these separate biases pulled in the same direction, ensuring that Republicans were heavily over-represented.

There are two ways in which pollsters try to eliminate bias. In *quota sampling*, the parent group itself is broken down by sex, age, income, religion, occupation, education, residential region, and any other characteristic relevant to the object of enquiry. Hair colour would have no bearing on a person's politics, for instance, but it would be important if we were studying attitudes towards a new cosmetic. Once the composition of the parent group is known, the pattern of a truly representative sample can be worked out by including similar proportions of the various categories. If the parent group has 60 per cent men and 40 per cent women, so must the sample, and so on. Questions are carefully prepared to ensure that they are not loaded and interviewers are trained to put them in an impartial manner. The interviewers are then sent out to find people to fit the sample. If their quota calls for ten, middle-income male factory-workers aged between 30 and 35, they must include ten, no more and no less. Experience soon teaches them how, when, and where to find the types they want. That is why street interviewers often ignore dozens of passers-by, then suddenly pounce. They have recognised a type needed to fill their quota, though of course they check before proceeding with the interview.

The trouble with quota sampling is that we may not have enough information about the parent group to construct a representative sample. Census returns would not tell us what proportion of high-income families in the north-west bought a used Ford as their second car. Yet this might be important in a car trade survey. The extent of introversion and extroversion among various groups may be important to researchers enquiring into the effects of smoking on health but the information does not exist.

Even when we know enough about the parent group to construct a model sample, the actual sample interviewed may not be as representative as we might wish. We can assume that interviewers are competent and that any who have yielded to the temptation of saving shoe-leather by filling in fictitious answers have been weeded out by spot-checks. We can never be sure, however, that they are not the victims of their own routines which lead them to look in particular places at particular times. To take a crude example, a survey of shopping habits might turn out seriously distorted results if interviewers found all their subjects outside city centre supermarkets. The behaviour of people who never shopped in supermarkets would be totally unrepresented.

For reasons like these, quota sampling has largely given way to *random sampling*. Now *random sampling* does *not* mean that subjects are chosen haphazardly. If they were, the results would be distorted still more by the interviewers' habits. One might spend all his time around bars; another might avoid them completely. A third might concentrate on the business section; another on a factory area. These again are extreme examples but we can see how easy it would be for a particular researcher's 'random' sample to be highly unrepresentative.

A truly random sample must be carefully selected, usually from a complete list of the parent group, for instance the register of electors. We can then form our sample by taking from it names at regular intervals, say every tenth, hundredth or thousandth, depending on how many we need. Better still we can use a table of random numbers which have been electronically generated and truly are random, meaning that each occurs with the same frequency as we should expect it would by pure chance and has been found to do so by exhaustive mathematical tests. The table then decides the intervals at which we should take names from our parent-group list.

Once the random sample has been compiled, interviewers track down the chosen subjects, calling on them many times if necessary. Repeat calls and the fact that some subjects live in remote places makes random sampling more expensive than quota sampling. Nor is it always practicable because a complete list of the parent group may not exist. When it is feasible, however, results tend to be more accurate than with quota sampling.

Just as important as the sample's composition is its size. It must be big enough to ensure that it is not biased by pure chance. Paradoxically, a seemingly large sample may be too small and a small one adequate. Everything depends on the complexity of the material and the relative frequency of the characteristic we are investigating. If a man from Mars took a truly random sample of ten humans and found that each had two legs, he would be right in assuming that the norm for humans was two legs per person. If he asked the same ten people about their literary tastes, it is unlikely that he would draw any meaningful conclusions because the range of possible answers greatly exceeds the number of people being questioned and if he asked whether they had a passionate interest in the music of Messiaen, he would discover nothing at all. Even a single 'yes' would suggest that far more of us were enthusiasts than is in fact the case and ten 'noes' would fail to reflect the tiny minority who do have a passionate interest.

These may seem elementary points but even professionals sometimes fail to appreciate them. In a test of polio vaccine, researchers divided a thousand children in one American town into two groups of 500, one of which was given the vaccine, the other not. When polio struck, all the vaccinated children escaped, and so did all those who had not been vaccinated. It was later realized that of *any* thousand children only 2 would normally be expected to catch polio during an epidemic. So only one child in either group was statistically at risk and the fact that no cases occurred in either group could be the result of chance, making the experiment worthless.[38]

Surprisingly, a sample of only 1500 is usually found to be adequate in a national opinion poll and it would have to be increased many times before the degree of accuracy was significantly improved.

There are, of course, numerous mathematical tests for checking the significance of sampling results. There have to be because there are so many different sampling situations. As Dr Russell Langley, an Australian medical statistician, pointed out, '... a test which will compare the performance of two rifles will not necessarily be suitable for comparing the performance of three rifles; a test which will compare accident rates in different factory shifts cannot be used for comparing the yields of a crop under the influence of different fertilizers. Tests vary, too, in the broadness of their applicability, in their accuracy, in their efficiency, and in the amount of arithmetic which they entail.'[39] The moral is clear. Anyone who needs to use sampling should call in an expert. Even then, he would be wise to take the results with a bucketful of salt.

7

The Fascination of Numbers

MAN is a pattern-making animal. He soon discovered that the
ten symbols which he had created were not just tools enabling
him to count and measure. They were the foundation stones of a
self-contained system with an infinity of themes and forms, most
of them of no practical value but so bewitching that they lured a
privileged few into a life-time's devotion. Hence was born the
profession of mathematics. Modern technology could not exist
without mathematics but we could reasonably argue that mathematics is primarily a game and that Concorde and Apollo are mere
spin-offs.

Alan Watts put it like this:

> 'The pure mathematician is much more of an artist than a
> scientist. He does not simply measure the world. He invents
> complex and playful patterns without the least regard for
> their practical applicability. He might almost be on a per-
> manent vacation—as if he were sitting on the terrace of a sea-
> side hotel, doing crossword puzzles and playing chess or
> poker with his cronies. But he works in a university, which
> makes it respectable.'[1]

Few of us will ever scale the Himalayas of mathematics but we
can all enjoy a stroll in the foothills where the dipping, twisting
pathways of our basic number system lead to a bonanza of
unexpected delights.

It seems, for instance, that we can take any number, reverse it,
add the two numbers together, continue to reverse and add, and
eventually end up with a palindrome, a number whose digits
appear in the same order whether they are read from left to right
or from right to left. With some numbers, it happens in a single
step.

e.g. 13.

13 + 31 = 44, which is a palindrome.

Other numbers may need two or more steps.

e.g. 64.

64 + 46 = 110

110 + 011 = 121, also a palindrome.

By the same process of repeatedly reversing and adding, 89 becomes 8,813,200,023,188 at the 24th step and 7,998 becomes 16,668,488,486,661 at the 20th. Of all the numbers under 10,000 only 249 fail to form palindromes in 100 steps or less and no number has yet been proved not to do so eventually.

Another series of palindromes is that formed by squaring numbers formed by the digit 1. We find that

$$1^2 = 1$$
$$11^2 = 121$$
$$111^2 = 12321$$
$$1111^2 = 1234321$$
$$11111^2 = 123454321$$
$$111111^2 = 12345654321$$
$$1111111^2 = 1234567654321$$
$$11111111^2 = 123456787654321$$
$$111111111^2 = 12345678987654321$$

Yet another palindrome is formed by multiplying 12,345,679 by 99,999,999. The product is 1,234,567,887,654,321.

As we range through the first few million numbers (the only ones that most of us are capable of comprehending) we find countless surprises. If we take any number of three digits in descending order, reverse it, subtract the new number from the original to get a third number, reverse it and add this fourth number to the third, the result is always 1,089.

e.g.   743
       − 347
       ‾‾‾‾‾
        396
       + 693
       ‾‾‾‾‾
       1089

Both 1,089 and some of its multiples are unusual for another reason. When multiplied, they yield a product which is the original number reversed.

e.g.　1089 × 9　= 9801
　　　2178 × 4　= 8712
　　　4356 × 1½ = 6534

When the numbers 312 and 221 are multiplied together, the product is the reverse of the product of themselves reversed.

　　　312 × 221 = 68952
　　　213 × 122 = 25986.

Numbers that recur in cycles are another curiosity. We can multiply 142,857 by any number between 1 and 6 and in each case, the product is the same series of digits in the same order but starting at a different point.

e.g.　142857 × 1 = 142857
　　　　　　 × 2 = 285714
　　　　　　 × 3 = 428571
　　　　　　 × 4 = 571428
　　　　　　 × 5 = 714285
　　　　　　 × 6 = 857142

Multiply it by 7 and the product is 999999.

When we work out the decimal fraction equivalent to one-seventh, we find the same digits recurring, 0·142857, and they appear yet again when we take the number 14, double it, move the product two places to the right, continue the process for as long as we wish and then add all the products together.

e.g.　14
　　　　28
　　　　　56
　　　　　112
　　　　　　224
　　　　　　　448
　　　　　　　　896
　　　　　　　　　1792
　　　　　　　　　　3584
　　　　　　　　　　　7168
　　　　　　　　　　　　14336
　　　————————————————————
　　　142857142857142857 . . .

Mathematicians look on isolated number problems as oddities. They can usually explain them on known principles and even when they can't, most find them far less compelling than the inexorable logic of an established law or the elegance of a simple solution to a complex problem. The appeal of mathematics has always been aesthetic, a fact we tend to overlook in these days of slide rules and pocket calculators. The ancients were less inhibited. They expressed their delight even in the examples they made up. The *Palatine Anthology*, a Greek collection probably dating from the fourth century A.D., is written in the form of poetry and includes a question based on the life of Diophantos, a celebrated number theorist who lived in Alexandria in the third century A.D. It runs:

> 'Here you see the tomb containing the remains of Diophantos, it is remarkable: artfully, it tells the measures of his life. The sixth part of his life God granted him for his youth. After a twelfth more his cheeks were bearded. After an additional seventh, he kindled the light of marriage, and in the fifth year he accepted a son. Elas, a dear but unfortunate child, half of his father he was and this was also the span a cruel fate granted it, and he consoled his grief in the remaining four years of his life. By this device of numbers, tell us the extent of his life.'[2]

(Answer: 84 years.)

The Hindus' approach to number was even more poetic. They used colours to represent unknown quantities and took their examples not from plumbers filling baths but from jewellers counting precious stones. 'These problems are stated merely for pleasure,' wrote Brahmagupta (588–660 A.D.) in one of his collections and some 200 years later, Mahaviracarya devised this conundrum for his *Ganita-Sara-Sangraha*:

> 'Into the bright and refreshing outskirts of a forest which were full of trees with their branches bent down with the weight of flowers and fruits, trees such as jambu trees, date-palms, hintala trees, palmyras, punnaga trees and mango trees —filled with the many sounds of crowds of parrots and cuckoos found near springs containing lotuses with bees

roaming around them—a number of travellers entered with joy.

'There were 63 equal heaps of plantain fruits put together and seven single fruits. These were divided equally among 23 travellers. Tell me the number of fruits in each heap.'[3] (One answer is 5.)

From the earliest times, mathematicians probed, scrutinized, teased, shuffled, manipulated, and categorized numbers, playing with them as children play with a box of bricks. They discovered, for instance, that some numbers were triangular, so-called because they can be represented by a series of triangles formed by dots. The apex consists of one dot and succeeding rows by two, three, four dots and so on. Each successive triangular number has an extra row of dots added at the bottom.

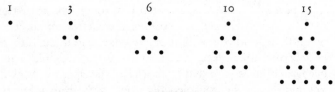

FIGURE I

When we look at these numbers more closely, we find that they share some unexpected properties. If we multiply any triangular number by 8 and add 1, we always get an odd number which is a perfect square.

e.g. $10 \times 8 = 80$ $80 + 1 = 81 = 9^2$
       $21 \times 8 = 168$ $168 + 1 = 169 = 13^2$

We can add together any two successive triangular numbers and the result is always a square. In fact, every square is the sum of two successive triangular numbers.

e.g. $1 + 3 = 4 = 2^2$
      $10 + 15 = 25 = 5^2$

Also, $9^2 = 81$, which is the sum of 36 and 45, both triangular numbers.

Numbers formed by adding successive even numbers are called oblong because they can be represented by oblong block of dots.

2	6	12	20	30
	(2+4)	(2+4+6)	(2+4+6+8)	(2+4+6+8+10)

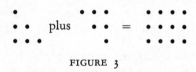

FIGURE 2

It will be seen that oblong numbers are exactly double the corresponding triangular numbers and can be represented by two triangles side by side.

$$
\begin{matrix} \bullet \\ \bullet\ \bullet \\ \bullet\ \bullet\ \bullet \end{matrix}
\quad \text{plus} \quad
\begin{matrix} \bullet\ \bullet\ \bullet \\ \bullet\ \bullet \\ \bullet \end{matrix}
\quad = \quad
\begin{matrix} \bullet\ \bullet\ \bullet\ \bullet \\ \bullet\ \bullet\ \bullet\ \bullet \\ \bullet\ \bullet\ \bullet\ \bullet \end{matrix}
$$

FIGURE 3

These oblong numbers also behave in some surprising ways. If we take them as a series (2, 6, 12, 20, 30, and so on), we find that each of them is equal to the square of the number representing its position in the series plus that number itself.

e.g. $2 = 1^2 + 1$ $12 = 3^2 + 3$
     $6 = 2^2 + 2$ $20 = 4^2 + 4$

It follows that we can build up the oblong series by taking every number in turn, starting with 2, squaring it and then subtracting the original number.

e.g. $2^2 - 2 = 2$ $5^2 - 5 = 20$
     $3^2 - 3 = 6$ $6^2 - 6 = 30$
     $4^2 - 4 = 12$ $7^2 - 7 = 42$

Squares are interesting quite apart from their relationship with triangular and oblong numbers. The square of any number higher than 1 is a multiple of 4 if even or a multiple of 8 plus 1 if odd.

e.g. Even $2^2 = 4 = 1 \times 4$ Odd $3^2 = 9 = 8 + 1$
$$4^2 = 16 = 4 \times 4$$
$$5^2 = 25 = (3 \times 8) + 1$$
$$6^2 = 36 = 9 \times 4$$
$$7^2 = 49 = (6 \times 8) + 1$$

Surprisingly, too, we can find the square of any number by taking the same number of consecutive odd numbers, starting with 1, and adding them up. Moreover, if we add 1 to the last of the five numbers, we shall find it is double the number we are squaring.

e.g. $4^2 = 1 + 3 + 5 + 7 = 16$ and $7 + 1 = 8 = 2 \times 4$
$6^2 = 1 + 3 + 5 + 7 + 9 + 11 = 36$ and $11 + 1 = 12 = 2 \times 6$

Squares of successive numbers also have an unexpected relationship. Their final digits form the palindrome 1, 4, 9, 6, 5, 6, 9, 4, 1 which is followed by a zero and then repeats itself indefinitely.

e.g. $1^2 = \underline{1}$ $4^2 = 1\underline{6}$ $7^2 = 4\underline{9}$ $10^2 = 10\underline{0}$
$2^2 = \underline{4}$ $5^2 = 2\underline{5}$ $8^2 = 6\underline{4}$ $11^2 = 12\underline{1}$
$3^2 = \underline{9}$ $6^2 = 3\underline{6}$ $9^2 = 8\underline{1}$ and so on

It can be seen that no squares ever end in 2, 3, 7, or 8. It can also be seen that the squares of any two numbers ending in digits which, added together, make 10, will each end in the same digit when squared.

e.g. $2^2 = 4$ $23^2 = 529$
$8^2 = 64$ $97^2 = 9409$

Squares can also be formed adding together the cubes of successive numbers starting from 1.

e.g. $1^3 + 2^3 + 3^3 + 4^3 = 1 + 8 + 27 + 64 = 100 = 10^2$

The root of a square so formed is equal to the sum of the roots of the cubes forming it.

e.g. $1 + 2 + 3 + 4 = 10$

Cubes have another relationship with successive odd numbers. Each is the sum of a series of these and the number of terms in the series is equivalent to its root.

e.g. $1^3 = 1 = 1$ (The root is 1 and there is one number in the
series)

$2^3 = 8 = 3 + 5$ (The root is 2 and there are two numbers in
the series)

$3^3 = 27 = 7 + 9 + 11$ (The root is 3 and there are three
numbers in the series)

There is a simple way of finding the first number of the series
required. The root is multiplied by itself less 1, and 1 is then
added to the product.

e.g. To find 6^3

$6 \times 5 = 30$

$30 + 1 = 31$

$31 + 33 + 35 + 37 + 39 + 41 = 216 = 6^3$

Oddly, every cube is a multiple of 7 or a multiple of 7 plus or
minus 1.

Problems relating to squares, cubes, and higher powers have
exercised the minds of some of the greatest mathematicians the
world has ever known, but for thousands of years they have also
appealed to ordinary people, even young children, as one of our
best-known nursery rhymes shows:

'As I was going to St Ives,
I met a man with seven wives,
Each wife had seven sacks,
Each sack had seven cats,
Each cat had seven kits,
Kits, cats, sacks and wives,
How many were going to St Ives?'

To the question as stated, the answer is 'none' because the
narrator himself was going to St Ives and since he *met* all the others,
they must have been coming away from it. But the trick ending has
almost certainly crept in at a late date, for the real interest of the
problem lies in the Chinese box aspect which, in mathematical
terms, means working out and adding together successive powers
of the number 7. As we shall see later, 7 has always had a special
significance for mankind and so, it seems, has the St Ives puzzle.
The *Papyrus Rhind* in the British Museum was written about

1800 B.C. in hieratic script by the Egyptian scribe Ahmes, who may well have been copying it from an earlier original. The final section is corrupt but a part of it reads:

'Houses	7
Cats	49
Mice	343
Ears of wheat	2,401
Hekat measure	16,807
Total	19,607'

Once again, we have successive powers of the number 7 and for many years, it was thought that the words 'houses', 'cats', 'mice', 'ears of wheat', and 'hekat measure' were not meant to be taken literally but were picturesque ways of representing squares, cubes, and so on. Later, however, the passage was compared with a section of Fibonacci's *Liber Abaci* published in A.D. 1202. One of the questions, which Fibonacci is thought to have taken from the common currency of the time, concerned seven old wives going to Rome. Each wife had seven mules, each mule had seven sacks, each sack had seven loaves, each loaf was accompanied by seven knives and each knife had seven sheaths. The problem was to work out how many wives, mules, sacks, loaves, knives, and sheaths were making the journey. The links with both the *Papyrus Rhind* and 'As I was going to St Ives' are obvious. When we look at all three examples, we find that wives, cats, sacks, and wheat (or loaves) occur in at least two of them. Astonishingly, a single number problem has continued to fascinate mankind for some 4,000 years and still delights each new generation of children.[4]

Number puzzles
Everybody likes number puzzles. We find them in all ages and most fall into one of three categories. The first, like the present-day version of 'As I was going to St Ives', hangs heavily on the wording of the question. Facts are not withheld but our attention is drawn away from some important element, rather as a conjuror's assistant distracts our attention while he palms an egg. These verses from a nineteenth-century magazine are a good example:

'Ten weary, footsore travellers,
  All in a woeful plight,
Sought shelter at a wayside inn
  One dark and stormy night.

"Nine rooms, no more," the landlord said,
  "Have I to offer you.
To each of eight a single bed,
  But the ninth must serve for two!"

A din arose. The troubled host
  Could only scratch his head,
For of those tired men no two
  Would occupy one bed.

The puzzled host was soon at ease—
  He was a clever man—
And so to please his guests devised
  This most ingenious plan.

In a room marked A two men were placed,
  The third was lodged in B,
The fourth to C was then assigned,
  The fifth retired to D.

In E the sixth he tucked away,
  In F the seventh man,
The eighth and ninth in G and H,
  And then to A he ran,

Wherein the host, as I have said,
  Had laid two travellers by;
Then taking one—the tenth and last—
  He lodged him safe in I.

Nine single rooms—a room for each—
  Were made to serve for ten;
And this it is that puzzles me
  And many wiser men.'[5]

How on earth can ten men be put into nine rooms and each have
one to himself? A closer look reveals the fallacy. The tenth man
was never in fact accommodated. Travellers 1 and 2 were put in
room A and it was one of these who was later given room I.
Traveller 10 was cleverly ignored.

The second type of puzzle is perhaps the most familiar and we
all came across it while still at school. 'Think of a number',
ordered some playground magician. We decided on, say, 12. He
went on, 'double it (24), add 10 (34), halve it (17), take away the
number you first thought of (17 – 12 = 5)—and the answer,' he
concluded triumphantly, 'is 5.' Bewildered, we wanted to know
how he did it. When he revealed that the answer was always half
the number added on, which could be any even number, we felt as
though we had been initiated into the Eleusinian mysteries.

Later, of course, we can work out for ourselves that he has
tricked us into eliminating everything brought in except for half
the added number and we can prove this by simple algebra. Most
'think of a number' problems can be similarly proved. If we ask
our subject to multiply his original number by 5, add 10, multiply
by 8, add 11, and multiply by 5, we have really manipulated him
into multiplying his original number by 200 and adding on 455.
So we can tell instantly from the result what number he originally
chose. We simply take away 455 and divide by 200.

```
e.g.   the number thought of is  17
         multiply by 5      =        85
         add 10             =        95
         multiply by 8      =       760
         add 11             =       771
         multiply by 5      =      3855
         Working back we find
       3855 – 455           =      3400
       3400 ÷ 200           =      17, the number first thought of.
```

We can make this type of puzzle still more mystifying by
juggling with several different numbers so that they appear in
separate columns in the answer. We could tell our subject, 'Take
the last three digits of your telephone number, double them,
multiply by 5, add 37, add the last digit of your street number,
multiply by 10, add the number of children in your family, and

take away 370.' Unless he has more than nine children, which is a risk worth taking, the first three digits of the result will be the last three of his telephone number, the next will be the last digit of his street number, and the final digit will be the number of children.

e.g. the last three digits of the telephone number are   779

double them	=	1558
multiply by 5	=	7790
add 37	=	7827
add last digit of street number, say 7	=	7834
multiply by 10	=	78340
add number of children, say 3	=	78343
take away 370	=	77973[6]

The third type of number problem is different again. It relies for its effect on number properties of which we are unaware. We may, for instance, be asked to choose any number between 1 and 31 and say on which of five cards it appears (Figure 4). The magician can identify the chosen number instantly by adding together the first numbers on the cards named. If the cards are B, C, and E, he adds together 16, 1, and 2 which gives him 19, the only number to appear on these three cards and no others.

A		B		C		D		E	
4	20	16	24	1	17	8	24	2	18
5	21	17	25	3	19	9	25	3	19
6	22	18	26	5	21	10	26	6	22
7	23	19	27	7	23	11	27	7	23
12	28	20	28	9	25	12	28	10	26
13	29	21	29	11	27	13	29	11	27
14	30	22	30	13	39	14	30	14	30
15	31	23	31	15	31	15	31	15	31

FIGURE 4

The secret is simple. The cards have been prepared by changing the numbers 1 to 31 from their decimal form into binary (see chapter 1), so that 2 is written 10, 3 becomes 11, 4 becomes 100 and so on. All the numbers, starting with 1, which in binary have the digit 1 in their right-hand column have been placed on one card (C), but in their original decimal form. Those with the digit 1 in the next column have been placed in decimal form on another card (E) and so on. By adding together the lowest numbers on the cards used, we are building up the chosen number bit by bit in its binary form. In our example, 16, 1, and 2 become 1000, 1 and 10 in binary, giving a binary total of 10011, and it will be seen that a number of this form can appear only on the cards named.[7]

In another trick depending on number properties, we are asked to write down a number of four or five digits and add them together. We must then strike out any digit we wish from the original number and subtract from the number formed by the remaining digits, the sum of the digits in the original. We must then add together the digits in the result and announce this final number to the magician. He immediately tells us what digit we struck out. He does this by mentally subtracting the stated number from 9. If it is greater than 9, he must first add its digits and then subtract.

e.g. The number thought of is 47352
    $4+7+3+5+2=21$
    Strike out, say 7 from 47352, giving 4352
    $4352 - 21 = 4331$
    $4+3+3+1=11$ (this is the number announced)
    The demonstrator adds 1 and $1 = 2$, subtracts 2 from 9
    and states the number struck out—7.[8]

The reason why the trick works becomes clear if we think out precisely what we are doing when we add the digits of a number together (or, in mathematical terms, find its digital root) and take the result from the number itself. The number 439, for instance, consists of $400 + 30 + 9$ and gives a digital root of $4 + 3 + 9$. When we take away the digital root, therefore, we are subtracting 4 units from the 4 hundreds, leaving four 99s, each of which is divisible by 9. Similarly, we are subtracting 3 units from the three 10s, leaving three 9s, each of which is also divisible by 9. We are also taking 9 units from the 9 units, leaving nothing. When we take away the digital root of any number, therefore, the result *must* be divisible by 9. Since we also know that the digital root of any number divisible by 9 is also 9, it must follow that the result of striking out one of the digits must be to reduce 9 itself by the value of that digit.

In passing, it is worth noting that 9 has many other intriguing properties, not because it is 'magic' but because it is one less than 10 in the decimal system.

The nine-times table (9, 18, 27, 36 etc.) has the digits 1 to 9 rising in the tens column and declining in the units, the reason being that at each step we are really adding 10 and taking away 1.[9]

From knowing that $9^2 = 81$, we can find the squares of other numbers consisting solely of 9s by adding an extra 9 to the beginning of each successive answer and inserting an extra 0 between the 8 and the 1.

e.g. $\quad 9^2 = 81$
$\quad\quad 99^2 = 9801$
$\quad\quad 999^2 = 998001$
$\quad\quad 9999^2 = 99980001$
$\quad\quad 99999^2 = 9999800001$

For cubes of numbers consisting of 9s, we add extra 9s to the beginning and end of the first number in the series—729—and insert extra 0s between the 7 and the 2.

e.g. $\quad 9^3 = 729$
$\quad\quad 99^3 = 970029$
$\quad\quad 999^3 = 997002999$
$\quad\quad 9999^3 = 999700029999$

Division of the first eight whole numbers by 9 also produces interesting patterns.

e.g. $1 \div 9 = 0\cdot1111$ recurring
$\quad 2 \div 9 = 0\cdot2222$ recurring
$\quad 3 \div 9 = 0\cdot3333$ recurring
$\quad$ and so on

Finally, if we take any number of three digits in descending order, reverse it and subtract the new number from the original, the new number will always be divisible by 9, its centre digit will itself be 9 and so will the sum of the other two digits.

e.g. $985 - 589 = 396$

Magic Squares

A magic square is simply a square divided into equal cells containing numbers arranged in such a way that any row, column or diagonal adds up to the same total. From early Babylonian times such squares have been thought to have magic powers and Geber, the eighth-century A.D. Muslim alchemist (it is not certain that he was a real person) believed that a square consisting of the first nine numbers held the key by which base metals could be turned into gold. In his engraving *Melancholy*, Albrecht Dürer (1471–1528) illustrates the popular custom of carrying a silver plate inscribed with a magic square as a talisman against the plague and as late as the nineteenth century, a Cairo seer used a blob of ink on a magic square as a miniature crystal ball to call up visions of people whom his clients wished to contact.

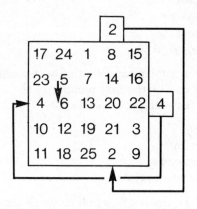

FIGURE 5

Magic squares can have an odd or an even number of cells and there are different methods of construction for various types. The simplest way of building an odd magic square is to place the number 1 in the centre cell of the top row. Consecutive numbers are then inserted diagonally upwards to the right and when this is impossible, according to the following rules:

(a) a number following one in the top row is placed at the bottom of the next column to the right;

(b) a number following one in the right-hand column is placed one row higher in the left-hand column;

(c) when the next cell is already occupied or the top right-hand cell is reached, the next number is inserted immediately below and the sequence continued as before. A glance at Figure 5 will show how this works out in practice.[10a]

Alternatively, we can build an odd square by placing the number 1 in the middle of the second row and as before work diagonally upwards to the right, following rules (a) and (b) where appropriate but ignoring (c) and instead inserting a blocked number two cells immediately *above*. When this takes us beyond the top row, we go back to one of the two bottom rows as in Figure 6.

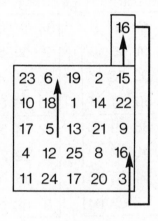

FIGURE 6

An even magic square of 16 cells can be constructed by drawing in the diagonals and inserting the numbers 1–16 consecutively (Figure 7a). Those falling on the diagonals are reversed and the square becomes magic (Figure 7b). A 64-celled square is slightly more complicated, for it has to be imagined as four separate 16-celled sub-squares. Besides the main diagonals, we have to draw in diagonals for the sub-squares. The cells are then numbered consecutively from 1 to 64 (Figure 8a). As before, numbers falling on the main diagonals are reversed but those falling on the sub-diagonals are first reversed and then transferred to the diagonal opposite (Figure 8b).[10b]

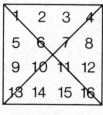

FIGURE 7a

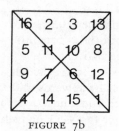

FIGURE 7b

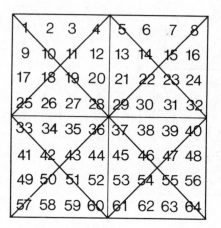

FIGURE 8a   FIGURE 8b

The 36-squared cell is different again. We imagine it as four 9-celled sub-squares, each of which is filled according to the rules for odd-numbered squares. The top left-hand sub-square has the numbers 1–9, the bottom right 10–18, the top right 19–27, and the bottom left 28–36 (Figure 9a). The main square becomes magic, however, only when we transpose the numbers 8, 5, and 4 in the top left sub-square and 35, 32, and 31 in the bottom left (Figure 9b). This square has at least two bonuses. The short diagonals 35, 32, 2+33, 5, 4 and 24, 23, 22+17, 14, 11 also add up to the same total as the rows, columns, and main diagonals.[10c]

(8)	1	6	26	19	24
3	(5)	7	21	23	25
(4)	9	2	22	27	20
(35)	28	33	17	10	15
30	(32)	34	12	14	16
(31)	36	29	13	18	11

35	1	6	26	19	24
3	32	7	21	23	25
31	9	2	22	27	20
8	28	33	17	10	15
30	5	35	12	14	16
4	36	29	13	18	11

FIGURE 9a   FIGURE 9b

Finally, it is worth taking a look at a bordered square, a magic square with a second magic square inside it. For a 25-celled bordered square, the inner square is constructed by adding 8 to the number in each cell of a basic 9-celled square, leaving the numbers 1–8 and 18–25 for the sixteen border cells. These are grouped into eight pairs, each totalling 26, and placed at opposite ends of the three middle columns, the three middle rows and the two diagonals (Figure 10).

1	2	19	20	23
18	16	9	14	8
21	11	13	15	5
22	12	17	10	4
3	24	7	6	25

FIGURE 10

There is no end to the making of magic squares. Some can be constructed according to still more sets of rules, others are built up by trial and error and may have an astonishing number of properties, which it is difficult, sometimes impossible, to explain. One of the most ingenious ever devised is that shown in Figure 11. The numbers in any row or column add up to 2056, as do those in any 4 × 4 block anywhere in the square, not to mention those in the four 2 × 2 blocks at the corners and those crossed by the marked chevrons. Moreover, any half column or row totals 1028. We can also take and number a chessboard in which each number is relative to its successor by a knight's move. Like other magic squares, it is useless but intriguing, It sets off an 'ah-hah' reaction that tingles through our body.

200	217	232	249	8	25	40	57	72	89	104	121	136	153	168	185
58	39	26	7	250	231	218	199	186	167	154	135	122	103	90	71
198	219	230	251	6	27	38	59	70	91	102	123	134	155	166	187
60	37	28	5	252	229	220	197	188	165	156	133	124	101	92	69
201	216	233	248	9	24	41	56	73	88	105	120	137	152	169	184
55	42	23	10	247	234	215	202	183	170	151	138	119	106	87	74
203	214	235	246	11	22	43	54	75	86	107	118	139	150	171	182
53	44	21	12	245	236	213	204	181	172	149	140	117	108	85	76
205	212	237	244	13	20	45	52	77	84	109	116	141	148	173	180
51	46	19	14	243	238	211	206	179	174	147	142	115	110	83	78
207	210	239	242	15	18	47	50	79	82	111	114	143	146	175	178
49	48	17	16	241	240	209	208	177	176	145	144	113	112	81	80
196	221	228	253	4	29	36	61	68	93	100	125	132	157	164	189
62	35	30	3	254	227	222	195	190	163	158	131	126	99	94	67
194	223	226	255	2	31	34	63	66	95	98	127	130	159	162	191
64	33	32	1	256	225	224	193	192	161	160	129	128	97	96	65

FIGURE 11

Amicable and Perfect Numbers

It is probably true to say that the making of magic squares has attracted more laymen than qualified mathematicians. Over the ages, serious number theorists have devoted much more time to the discovery of amicable and perfect numbers. Amicable numbers are so called because each of the pair is equal to the sum of the other's divisors, excluding itself.

e.g. 220 and 284
> The divisors of 220 are 1, 2, 4, 5, 10, 11, 20, 22, 44, 55 and 110, which add up to 284
> The divisors of 284 are 1, 2, 4, 71 and 142 which add up to 220

This pair, the smallest known, was familiar to the ancients but the next smallest—1184 and 1210—was discovered as late as 1866 by the Italian Nicolo Paganini when he was still only sixteen. Leonhard Euler, an eighteenth-century Swiss mathematician, published more than sixty, including 17,296 and 18,416, and 9,363,584 and 9,437,056, pairs which had been discovered respectively by Samuel Fermat in 1636 and by René Descartes in 1638. By then, complicated rules for discovering amicable pairs had been formulated and proved and by 1926, 390 had been listed.

The search for perfect numbers has also occupied mathematicians for at least 2,000 years. They are called perfect because they are equal to the sum of their divisors, excluding themselves.

e.g. 6
> Its divisors are 3, 2, 1
> $3 + 2 + 1 = 6$
> Therefore 6 is perfect

Euclid discovered that perfect numbers were linked with a form of prime number later known as Mersenne primes after the seventeenth-century mathematician Marin Mersenne who studied them at length. Every Mersenne prime is formed by a power of 2, minus 1.

e.g. $2^2 - 1 = \quad 3$, a Mersenne prime
$\quad\; 2^3 - 1 = \quad 7$, a Mersenne prime
$\quad\; 2^5 - 1 = \;\; 31$, a Mersenne prime
$\quad\; 2^7 - 1 = 127$, a Mersenne prime

Not all numbers generated in this way are prime, however.

e.g. $2^4 - 1 = \quad 15$, which has the factors 3 and 5
$\quad\quad 2^{11} - 1 = 2047$, which has the factors 23 and 89

Euclid proved that a number could be perfect only if it took the form $2^{p-1}(2^p - 1)$ and $2^p - 1$ itself was a Mersenne prime. The search for Mersenne primes and perfect numbers went hand in hand, and in 1811 Peter Barlow of Britain's Royal Military Academy listed the first eight perfect numbers, the only ones then known. The greatest of these was 2,305,843,008,139,952,128 formed from the Mersenne prime derived from 2^{31}. It was, he said, 'probably the greatest that ever will be discovered; for, as they are merely curious without being useful, it is not likely that any person will attempt to find one beyond it.'[11] However, he underestimated the pertinacity of his colleagues. By 1914, another four Mersenne primes had been discovered and since then computers have thrown up twelve more. The largest now known is $(2^{19937} - 1)$, giving the perfect number $(2^{19937} - 1) \times 2^{19936}$, a notation which may irritate non-mathematicians but saves several pages of print. Written in full, it has 12,003 digits.[12]

Like all the rest of the patterns and problems we have been considering in this chapter, perfect numbers and amicable pairs are simply phenomena of the system created by man. They were studied by generations of mathematicians simply for their inherent interest. They were not magic. The only power they exercised was that by which an elegant and obstinate puzzle casts a spell over the human mind. In the beginning, however, they were called 'amicable' and 'perfect' because it was believed that they did indeed have power over, or at least corresponded with, events in the material world. In the fourteenth century, the Arab scholar Ibn Khaldun had this to say of amicable numbers:

> 'Persons who occupy themselves with talismans assure that these numbers have a particular influence in establishing union and friendship between two individuals. One prepares a horoscope theme for each individual, the first under the sign of Venus while this planet is in its house or in its exaltation and while it presents in regard to the moon an aspect of love and benevolence. In the second theme the

ascendant should be in the seventh sign. On each one of these themes one inscribes one of the numbers just indicated, but giving the strongest number to the person whose friendship one wishes to gain. I don't know if by the strongest number one wishes to designate the greatest one or the one which has the greatest number of parts. There results a bond so close between the two persons that they cannot be separated.'[13]

Perfect numbers were long venerated because the first two were 6 and 28. It was known that God had created the world in 6 days and that it took the moon 28 days to travel round the earth. It must follow, therefore, that other perfect numbers, when discovered, would reveal, or reflect further truths about the universe.

Whether or not we accept this view, new developments in mathematics continue to throw light on problems ranging from the esoteric to the commonplace. The results, though, are couched in language so specialized that they are often beyond the understanding of laymen. Discussing the mediaeval poser 'How many angels can sit on the point of a needle?', Professor H. H. Rosenbrock wrote:

'If angels are located at mathematical points, then the question becomes a debate about set theory. In medieval times this was no doubt a subject filled with paradox. In modern times, mathematicians have settled upon definitive answers in which they all agree.

'So for example . . . any countably infinite set of angels can sit upon the point of a needle, but not every uncountably infinite set. Similarly, it is possible to define a space which is connected but not locally connected, in which an angel can dance from any point to any other, but cannot dance, in a continuous manner, in a circle around any point.'[14]

In comparison, it might seem simple to prove that a cartographer never needs more than four colours to draw a map in which every country is coloured differently from its neighbours. Yet mathematicians have been struggling to do so since at least 1852. Success finally came to Professors Kenneth Appel and

Wolfgang Haken after four years of research at the University of Illinois. They clocked up a thousand hours on a computer and drew on the four-foot pile of print-out for 700 pages of additional material to bolster the 200 pages of their proof. Professor Appel explained:

> 'It was terribly tedious with no intellectual stimulation. There is no simple elegant answer, and we had to make an absolutely horrendous case analysis of every possibility. I hope it will encourage mathematicians to realize that there are some problems still to be solved, where there is no simple God-given answer, and which can only be solved by this kind of detailed work. Some people might think they're best left unsolved.'[15]

It seems indeed that many problems may well remain unsolved simply because they are insoluble by their very nature. Professor Kurt Gödel of the Institute of Advanced Study at Princeton, New Jersey, proved that nothing in arithmetic is certain, not even within the system itself. *Internal* contradictions may appear at any time because there are always problems which cannot be proved or disproved from the axioms we started with. The more we explore, the more we have yet to find. So number theory is a maze of endless corridors: every time we push open a door, we find another door ahead. For mathematicians, that is the most delightful prospect of all.

8

Divine Proportion

THERE is a harmony in the universe and it can be expressed in terms of number.

If you take a daisy and look closely at the golden pin-cushion of its disc, you will find that it consists of dozens of tiny, tightly packed individual flowers or florets. These are not arranged haphazardly but in two sets of curved lines spiralling out from the centre. Twenty-one of them spiral in a clockwise direction, the other thirty-four in an anti-clockwise direction.

The numbers 21 and 34 are not haphazard either. More than seven hundred years ago, the mathematician Leonardo of Pisa (Fibonacci) included in his *Liber Abaci* the following problem:

> 'A certain man put a pair of rabbits in a place surrounded on all sides by a wall. How many pairs of rabbits can be produced from that pair in a year if it is supposed that every month each pair begets a new pair which from the second month on become productive?'[1]

We soon see that in each of the first and second months only 1 pair will be born but as the later pairs start producing, we get 2 pairs in the third month, 3 in the fourth, 5 in the fifth, 8 in the sixth, 13 in the seventh and so on. In other words, the number of pairs produced in each successive month is equal to the sum of the pairs produced in the two previous months. This sequence, the first of its kind discovered in the West, became known as the Fibonacci series and it will be seen that its eight and ninth terms are 21 and 34. We have, therefore, a direct link between a daisy and a seemingly arbitrary pattern of numbers worked out millions of years after daisies first evolved and hundreds of years before anyone studied their structure.

We cannot write it off as a coincidence because other plants also have links with the Fibonacci series. The scales of pine-cones form a similar pattern of criss-crossing spirals, 5 of them running clockwise, 8 anti-clockwise. The knobbles on pineapples run in 8 clockwise spirals and 13 anti-clockwise. The leaves of many plants successively join their stems at angles increasing at the same proportion as the Fibonacci series, (Figure 1), which is also related to the structure of animal horns and the genealogy of the male bee. Californian mathematicians were so fascinated by these links that in 1962 they formed a Fibonacci Association to 'exchange ideas and stimulate research on the Fibonacci numbers and related topics.'[2]

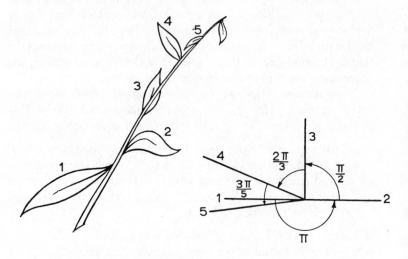

FIGURE 1 The angles of leaves on a stem

Time and again we find in nature forms and relationships which mathematicians thought they had invented as purely theoretical concepts: the cube-shaped crystals of common salt, the varied hexagons of snow-flakes, the shell of the nautilus mollusc whose chambers follow a logarithmic spiral—that is, a spiral which forms exactly the same angle with any radius drawn from its centre to any point along its length.

These links between nature and number led Dr Tobias Danzig, Professor of Mathematics at the University of Maryland, to compare the mathematician with 'a designer of garments, who is utterly oblivious of the creatures whom his garments may fit.' He continued:

'There have been quite a few such delightful surprises. The conic sections, invented in an attempt to solve the problem of doubling the altar of an oracle, ended by becoming the orbits followed by the planets in their courses about the sun. The imaginary magnitudes invented by Cardan and Bombelli [sixteenth-century mathematicians] describe in some strange way the characteristic features of alternating currents. The absolute differential calculus, which originated in a fantasy of Riemann [a nineteenth-century mathematician] became the mathematical vehicle for the theory of Relativity. And the matrices which were a complete abstraction in the days of Cayley and Sylvester [nineteenth-century mathematicians] appear admirably adapted to the exotic situation exhibited by the quantum theory of the atom.'[3]

In 1864, the English industrial chemist John Newlands made another odd discovery. If the elements then known were listed according to their atomic weights, each shared properties with those occuring eight places above and eight places below. Contemporaries ridiculed his 'law of octaves' because each element was thought to be unique and totally unlike the rest. Classifying elements by their atomic weights made as much sense, said the scoffers, as listing them in alphabetical order.

The Russian chemist Dmitry Mendeleyev was working on the same lines as Newlands and five years later drew up the first *periodic table*. This also showed relationships between different elements occurring at regular intervals in the scale of atomic weights. The objectors pointed to discrepancies. Mendeleyev checked and found that the atomic weights of seventeen elements had been wrongly calculated. When these were corrected, the discrepancies disappeared. Moreover, gaps in the table enabled Mendeleyev accurately to predict the existence and even the properties of elements that had not yet been discovered.

As modern physics developed, it became clear why Mendeleyev had been right. It was found that the nucleus of each atom was surrounded by electrons arranged in one or more of seven possible shells or layers, rather like onion-skins. In the present-day version of the periodic table, all the elements in the horizontal line (or period) numbered 1 have two electrons, while those in period 2 all have eight, as do those in period 3. The elements of periods 4 and 5 have eighteen electrons and those in period 6, thirty-two. Those in the incomplete seventh period also have thirty-two. In addition, the elements in each vertical column have the same number of electrons in their outer shells, giving them the same power of combining with other elements. This is why they are broadly similar to each other in acidity, hardness, melting-point, and other physical and chemical properties. (Figure 2.)

New number patterns are being discovered in nature all the time. In 1967, the Swiss scientist Hans Jenny published his book *Cymatics*, a study of the effects of vibration on various substances. Each behaves in a different way. If we toss a handful of sand onto a circular plate vibrating at 1060 cycles per second, it forms a 'flower' of four petals converging on a centre. Sand always makes this same four-petalled flower but thin liquids run into a ring of eight separate patches corresponding to the sand petals and the spaces between them. Thicker liquids behave differently again. Hot kaolin first forms circular wave patterns, one of which has a dip in the middle. As the kaolin cools and becomes thicker, the dip turns into a bulge which grows into a solid dome with a regular number of wave patterns spiralling up the sides and over the top, rather like a jelly that has been turned out of a mould. The process continues and by the time the kaolin has set, it looks more like a tree ringed with stunted branches.[4]

Once scientists start experimenting with the patterns found in nature, the question arises, 'Are they practising science or art?' And we could ask the same question of some artists. Computers can be programmed to compose music, choreograph dances and write poems or essays, either wholly or partly at random. They can be linked with plotters to draw the most exquisite abstract forms. Peter Milojević, a Yugoslav who settled in Canada, used an IBM 7044 computer and a Calcomp plotter to draw 'trees' which threw out branches at intervals based on the Fibonacci series. By programming the computer to produce lines of different textures, he created designs that were natural, elegant, and even poetic.[5]

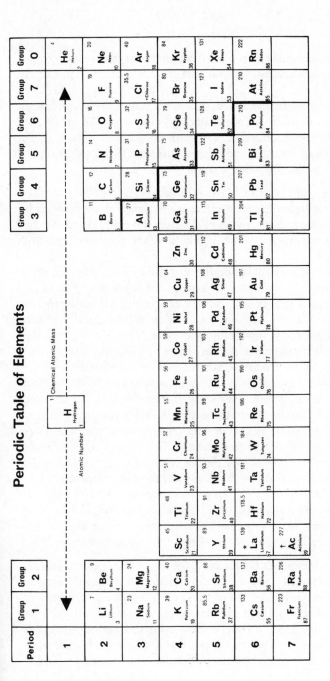

Periodic Table of Elements

FIGURE 2

Tsai Wen-ying has also used number to create his 'cybernetic sculptures'. After an early training in brushwork at his home town of Amoy, China, he studied mechanical engineering at the University of Michigan and had a successful career in the American construction industry before changing back to art. An exhibition of his work was described as 'a total environment' by the English critic Jonathan Benthall who specializes in the application of technology to art. Benthall wrote:

> 'The overall effect is from bluish high-frequency strobes [flashing lights], which create a strange and other-worldly visual effect. Each piece is different but they have certain design features in common, like different pieces of the same genus.
>
> 'Each consists of a number of stainless-steel rods set on a platform vibrating at a consistent and unvarying rate of 20 to 30 cycles per second. But the flashing of the strobe makes the eye see the rods as oscillating asymmetrically. Each flash lasts for a few millionths of a second only, and the intervals between the flashes are of variable duration. When the rate of the flashes equals the rate of the vibrations of the rods—we may call this the "synchronous rate"—the motion of the rods appears stationary in the shape of a harmonic curve. When the rate of the strobe-flashes is altered to slightly slower or faster than the rate of vibration, then the rods appear to be slowly undulating. The greater the deviation between the rate of the flashes and the constant harmonic motion of the rods, the more rapidly the rods appear to move. There is thus a range from relaxed undulation to excited palpitating. The direction of the apparent spiralling (clock-wise or counter-clockwise) depends on whether the rate of the strobe-flashing is above or below the synchronous rate.'[6]

Tsai's work depends for its effect on the response of the human eye to rods vibrating at a set rate in the light of strobes flashing at regular but varying rates. All of these rates are measured numerically and by changing them, we can change the image perceived by viewers' brains. Is an 'art' dependent on mathematical principles really art? Romantics would say not but if we counted heads over the ages, at least in Western civilization, the romantics would

probably be in a small minority. The majority opinion has always
been that number has an important part to play in almost any art.

Among the Greeks, Plato held that the highest form of beauty
was that expressed in mathematical terms. In the *Philebus*, Socrates
argues:

> 'By beauty of form, I do not mean, as is commonly meant,
> the creatures of nature and pictorial art. But let us put it like
> this: I mean straights and curves and all that a lathe, rule or
> square may produce from them in plane or solid form . . .
> These, I maintain, are not instances of relative beauty, like
> other things, but are eternally and essentially beautiful.'

Aristotle took a similar line in his *Metaphysics*. He wrote:

> 'Those who claim that the mathematical sciences are not
> concerned with goodness and beauty miss the truth. For
> mathematics pronounces and demonstrates on such matters
> in a marked degree. True, it may not deal with concrete
> embodiments, but it is by no means silent on underlying
> principles. For instance, the greatest species of beauty are
> order, proportion and limit, which are above all the objects of
> mathematical research.'[7]

Order, proportion, limit: we often take them for granted but
when we look around, it is obvious that they are intrinsic in so
many forms of art—the structure of a sonnet, the ingredients of
a recipe, the figures of a dance. At Abydos in Egypt, we can still
see in the fourteenth-century B.C. temple of Seti I, a relief in
which human figures follow exactly the proportions of the
Fibonacci series in the distances from the upper border to the tops
of their heads, from the tops of their heads to their waists and from
their waists to their feet. In his book *The Curves of Life* (1914) Sir
Theodore Cook showed that a Botticello Venus also had Fibonacci
proportions.

Modern western music is highly dependent on number both in
its notation and structure. This is obvious enough in the works of
Baroque composers but even in the continuous flow of the Prelude
to Wagner's *Tristan and Isolde* there are three repetitions of the
phrase that occurs in bars 17–21 with exactly the same number of
bars between each one and the next. Bruckner too relied heavily

on numerical order, for instance, in his heavily revised Symphony No. 8 first performed in Vienna in 1892. Phrases of 2, 4, 6, and 8 bars in the opening section of the *Scherzo* build into a double sequence of 32 + 32 bars. In 1925, Berg wrote his *Concerto for Piano, Violin and thirteen Wind Instruments* and based it almost entirely on the number 3 and its multiples. He wrote in a letter: '. . . I realise that—in so far as I make this generally known—my reputation as a mathematician will grow in proportion (. . . to the square of the distance) as my reputation as a composer sinks.'[8]

More recently, experimental composers have also used number extensively, though 'order, proportion and limit' are probably the last things they intended. In 1951, while writing his *Sixteen Dances*, the American John Cage arranged sections of material in the form of magic squares which he linked together by throwing *I Ching*, the traditional Chinese yarrow-stick oracle. He believes that silence is as important as sound in music and his piece *4' 33"* consists of four minutes and thirty-three seconds of total silence. Yet the score is divided into parts I, II, and III, all with the direction *tacet*.[9]

Perhaps the most familiar examples of the application of number to art, however, are the great Gothic cathedrals of western Europe. Architectural historians still argue about the rules applied or even the existence of an overall scheme of proportion but all agree that mathematics played a large part. The main axes often

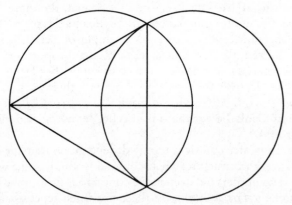

FIGURE 3 Vesica piscis. Euclidean figure resembling the fish symbol.

corresponded with those of the *vesica piscis,* a Euclidean figure that became important to Christians because of its resemblance to the key symbol of the fish. It is formed by the intersecting arcs of two equal circles so that the vertical axis of the resulting pointed oval or *vesica* forms two equilateral triangles when its ends are joined to the far ends of the diameters of each circle (Figure 3).

The cathedral was then divided into equal bays and the heights of the various parts decided by the heights of equilateral triangles based on the floor beneath. Rarely, if ever, was this system consistently applied but we have only to look at the star polygons in chapter-house plans and in the elaborate tracery of vaulted ceilings to realize that Gothic architects had a sound working knowledge of mathematics. (Figures 4 and 5.)

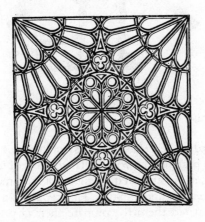

FIGURE 4 Vaulted ceiling—Dean's chapel, Canterbury Cathedral.

Why the Gothic style should have given way to the Renaissance during the fifteenth century is not altogether clear. The new style was largely pioneered by Filippo Brunelleschi (1377–1446), a Florentine goldsmith and sculptor, who later turned to engineering and architecture and made his name with the dome he built for Florence cathedral. Though still largely Gothic in style, the dome has many classical features which link it with his other masterpieces, for instance the exquisitely arcaded Foundling Hospital in Florence and such churches as Santo Spirito which Rupert

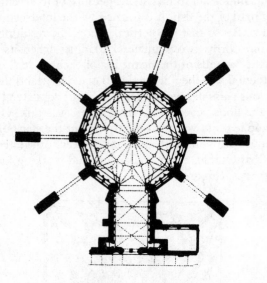

FIGURE 5 Ground plan—chapter house of Lincoln Cathedral

Furneaux Jordan described as 'a symphony of semi-circles'.[10] Brunelleschi's practical approach, his lightness of touch and above all his classicism, exactly fitted the mood of the times. Florence had grown rich on trade and banking. Her people had acquired a self-confidence that led them to compare their city with Rome and even Athens, giving them a zest for art and learning that received an added stimulus when refugee scholars poured in from Constantinople after its sack by the Turks in 1453. In their architecture, it was natural that they should turn for inspiration to the works of their Roman predecessors, rather than to the Church-dominated Gothic of northern Europe. Besides, they found in classical literature architectural theories firmly grounded in mathematics. These applied twice over to men who worshipped reason and money with equal fervour.

The working out of these ideas during the next hundred years provided the focal point of one of the most delectable views in the world, that of San Giorgio Maggiore seen across the lagoon from the Doges' Palace in Venice. Henry James called it, 'a success beyond all reason', and added: 'It is a success of position, of colour, of the immense detached campanile tipped with a golden angel'.[11] It was designed by the most influential architect of the late Renaissance, Andrea di Pietro (1508–80), who, as a 28-year-old stone-mason, was hired by the humanist poet Giangiorgio Trissino to work on his villa near Vicenza. Trissino saw possibilities in Andrea, made him his protégé and eventually renamed him after a leading figure in one of his poems—Palladio, guardian angel of the sixth-century general Belisarius who drove the Goths from Italy and so preserved the classical tradition.

The new Palladio became an architect who crowned the classical revival with a style widely copied in western Europe. In England, it was taken up by Inigo Jones in the seventeenth century and again by Lord Burlington, William Kent, and others in the eighteenth century. Palladian façades with the pediment and columns of a classical temple portico became one of the best-known features of the English landscape. In America, Thomas Jefferson chose the Palladian style for Monticello, the home he started to build for himself in 1770 at Charlottesville, Virginia.

Like other Renaissance architects, Palladio believed that the cosmos was God's creation and therefore perfect. Each part was related to the others and also to the whole by a divine harmony which could be measured mathematically. Architects must ensure a reflection of this harmony in their buildings:

> 'Beauty will result from the most beautiful form and from the correspondence of the whole to the parts; of the parts among themselves, and of these again to the whole; so that the structures may appear an entire and complete body, wherein each member agrees with the others and all members are necessary for the accomplishment of the building.'[12]

If man was made in the image of God, it seemed reasonable to expect that the divine harmony could be traced in the human body. Vitruvius, the first-century B.C. Roman architect had apparently taken this view, for he showed how a man could be fitted into both a square and a circle, both figures of perfection. Moreover, his legs and an imaginary line joining his feet formed an equilateral triangle. (Figure 6.)

Vitruvius had also apparently worked out the proportions of his model man. I say 'apparently' because this chapter of his book *De Architectura* is far from clear. True, attempts had been made to show how mediaeval cathedrals were planned round the shape of the human body with the apse corresponding to the head, and in

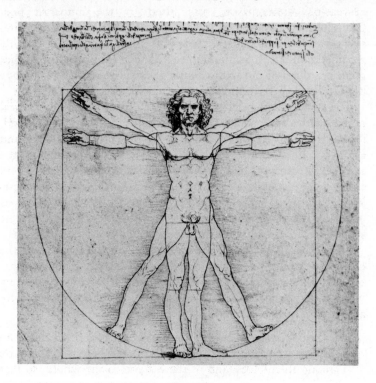

FIGURE 6 Vitruvian man

the sixteenth century Leonardo da Vinci had worked out the proportions of a well-made man, fitting him into a circle with his navel at the centre. Yet it is doubtful if any living man ever had these proportions and on the whole it seems unlikely that Vitruvius intended any such implication. More probably, he was using an imaginary human figure to demonstrate a *system* of proportion. Broadly speaking, this was based on a basic unit or module in multiples of which every dimension could be expressed.

These multiples, however, had to be exactly divisible into the number of parts making up the whole. If we had a column of 96 units, its various parts must not measure 5, 7, 9, 10, 11, or 13 units (and so on), because these do not divide exactly into 96. The preferred proportions would be:

1	2	4	8	16	32
3	6	12	24	48	96

These are related to each other through the basic unit and also to the whole as simple fractions:

1/96	1/48	1/24	1/12	1/6	1/3
1/32	1/16	1/8	1/4	1/2	1

According to this interpretation of Vitruvius, therefore, we have an arbitrary system of proportion based on geometric progressions unrelated to the human body.[13]

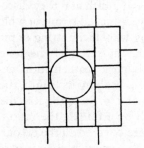

FIGURE 7 Villa Rotonda—plan

Palladio used the Vitruvian system and stuck to a rigid symmetry of which his Villa Rotonda is probably the most familiar example. (Figure 7.) If we fold the plan of almost any of his villas down the centre, we find that every room or corridor on one side fits exactly over a corresponding room or corridor on the other. For the actual proportions, he followed the 'musical analogy' put forward by the poet, scholar, and architect Leon Battista Alberti (1404–72). In his *Ten Books on Architecture*, Alberti wrote:

'I am every day more and more convinced of the truth of Pythagoras' saying, that Nature is sure to act consistently and with a constant analogy in all her operations: from whence I conclude that the same numbers, by means of which the agreement of sounds affects our ears with delight, are the very same which please our eyes and our mind. We shall therefore borrow all our rules for the finishing of our proportions from the musicians, who are the greatest masters of this sort of numbers.[14]

Almost all the architects of the Renaissance danced to the musical numbers of Pythagoras and his followers. Since both music and mathematics were traditionally 'liberal' arts to which practitioners of such 'mechanical' arts as painting, sculpture, and architecture aspired, the adoption of musical harmonies had the double advantage of raising the status of architecture and of rooting it firmly in the classical tradition. But why were numbers linked with music in the first place?

The Greeks preferred musical consonances based on intervals of an octave, a fourth, and a fifth and they discovered that the pitch of a note given off by a plucked string or a blown pipe varied in a particularly intriguing way. If the string or pipe was stopped half-way along its length, the pitch of the note given off rose by exactly an octave, giving a mathematical ratio of 1:2. If it was stopped a third of the way along its length, the pitch would rise by a fifth, giving a ratio of 2:3. Stopping it at the three-quarter mark would raise the pitch by a fourth, the ratio then being 3:4. The ratios 1:2:3:4 therefore covered all the favoured consonances and could be extended to the octave-plus-fourth, octave-plus-fifth, and double-octave. As with the daisy spirals, a purely natural phenomenon had been found to follow precisely a man-made number system. As Wittkower points out, 'One can understand that this staggering discovery made people believe that they had seized upon the mysterious harmony that pervades the universe.'[15]

The architects of the Renaissance believed profoundly in this 'mysterious harmony' and had no difficulty in making the imaginative leap from music to building. We can see how they applied the analogy when we look at the mediaeval church of Santa Maria Novella in Florence. In the middle of the fifteenth century, Alberti designed for it a new façade which has been called 'the first great Renaissance exponent of classical *eurhythmia*.'[16] (Figure 8.) The façade as a whole fits into a perfect square

which is divided equally into two storeys so that we have an overall ratio of 1:2. The square enclosing the main part of the upper
storey is exactly half as wide as the lower storey giving the same
1:2 ratio. This occurs yet again in the relationship between the
height of the upper pilasters and the height of the upper storey as
a whole and also in the relationship between the centre bay and the
total width (excluding the side scrolls). The entrance bay of the
lower storey has the same width as the central bay of the upper
storey making it a quarter of the total width of the lower storey
(1:4) but is one-and-a-half times as high, giving a width-to-height
ratio of 2:3. The same ratio is echoed in the proportions of the
door panels. Decorative squares on the attic (only one of these is
shown) are a third of its height (1:3) and twice the diameter of the
upper columns (2:1). Almost all the parts of the façade are therefore related to themselves and to the whole in ratios forming the
basic Pythagorean series 1:2:3:4.[17]

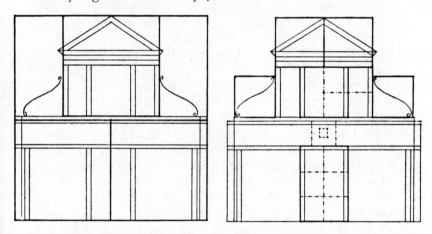

FIGURE 8 The façade of Santa Maria Novella

Palladio's proportions tended to be more complicated. He had
three separate ways of fixing the height of a room, all of them
forming a mean between its length and breadth. It could be an
arithmetical mean with the same measurement separating the
height from the other dimensions, for instance a 6ft × 12ft room
would be 9ft high, the difference between height and the other two

dimensions being 3ft. Or it could be a geometric mean with the height having the same proportion to the breadth as to the length, so a 4ft × 9ft room would be 6ft high (4:6 = 6:9). Alternatively, it could be harmonic, which meant that the fraction formed by dividing the difference between breadth and height by the breadth itself was the same as the fraction formed by dividing the difference between height and length by the length. The height of a 6ft × 12ft room would thus be 8ft, since $\dfrac{8-6}{6} = \dfrac{12-8}{12}$.[17a]

Palladio following Alberti also used extended musical harmonies. The ratio between the breadth of a room and its length was commonly 2:3, which is the same as 4:6. If we extend the length (6) by the overall ratio 2:3, we add to it a further 3 units making 9 in all. The extended ratio for the room is then 4:9. Similarly, the proportion 9:16 could be derived from a basic 3:4, which is the same as 9:12, by adding to the 12 a number related to it in the same ratio as 12 is to 9, i.e. 4, making 12 + 4 = 16. Even more complicated sub-ratios were devised but they were not to be used indiscriminately for this would destroy the underlying harmony. If a particular sub-ratio was used in one part of a building, the sub-ratios used in the rest must belong to the same 'family'.[18]

We can see how Palladio used these various proportions when we look at the plans he drew for his book *Quattro libri dell' architettura* which appeared in 1570. He often used the simple ratios 1:1, 1:2, 2:2, and 2:2, 2:3, 3:3 as in the Villa Thiene at Cicogna and in the Palazzo Porto-Colleoni. The Villa Godi Porto, started at Lonedo around 1540, has a central portico leading into a large hall with four rooms on either side, each measuring 16ft × 24ft (2:3), as does the portico. The hall of 24ft × 36ft has similar proportions and if we take both measurements together, we find the sequence 16, 24, 36, which gives the extended ratio 4:6:9. Musically it could be thought of as two successive fifths. (Figure 9.)

Other villas show variations on the same theme, though ratios of 3:5, 4:5, 5:6 and 5:9 also creep in. Palladio was not abandoning first principles but taking into account the fact that musical theory had itself developed during the sixteenth century and intervals such as major thirds, major and minor tenths and elevenths were added to the Pythagorean consonances. Since these were a further revelation of cosmic harmony, it was legitimate, even essential, to echo them in architecture.

It may be thought that the system was now so complicated that we are reading into it something that was never intended. We have documentary evidence to the contrary. Although practical difficulties sometimes forced Palladio to make changes when it came

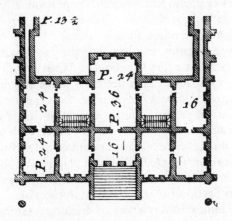

FIGURE 9 Plan of Villa Godi Porto

to building, we have his original plans in his *Quattro libri* and these follow the musical analogy closely. One of his most important clients was Daniele Barbaro who wrote a commentary on Vitruvius, incidentally supporting the same theory, which was also insisted on by the monk Francesco Giorgio in his *De harmonia mundi totius* (1525). Wittkower concludes: 'Nobody can deny that Palladio's numbers were meant to be indicative of certain ratios and it is not this fact but only the degree of interpretation which may be questioned.'[19]

The reaction against Renaissance ideas of proportion did not come until the eighteenth century, when they were attacked from all sides. Hogarth asked bluntly what music had to do with the visual arts. With enthusiasm for the classics no longer white-hot, no one could find a convincing answer. Other critics pointed out that the beauty of a building lay not in the architect's plan but in what the viewer saw. As he walked around, staring at different parts from different angles, he could not possibly appreciate the subtleties of 4:6:9 and even more complicated ratios, and still less their relationship to the whole which he was never able to see. The Romantics believed that beauty lay in the eye of the beholder, not in the proportions of the beheld, and Ruskin too came out against 'finite rules' of proportion. It was, he said, 'a matter of feeling and experience.' At the same time, he broadly supported the stand of Bishop Berkeley who had written: 'Proportions are to be esteemed just and true, only as they are relative to some certain use or end', thus anticipating modern functionalism.²⁰

In practice, feeling and function proved treacherous guides. A Brunelleschi or a Sir Christopher Wren could safely rely on his intuition but lesser men needed a set of standards that were generally agreed. Fitness for purpose was no substitute for these standards: an office block might provide efficient and comfortable working conditions yet be an eyesore. Faced with the nineteenth-century building boom, architects rummaged through the history of aesthetics for a practical system. Hogarth's 'line of beauty' based on the cone was one possibility but perhaps that based on the Golden Section proved the most influential.

The Golden Section is a way of dividing a segment so that the smaller part is to the greater as the greater is to the whole. Mathematically, it was known to the Greeks but there is no evidence that they *consciously* used it in art or architecture. Lucas Pacioli, the fifteenth-century Tuscan monk and mathematician, seems to have been the first man to point out its aesthetic possibilities. It was, he said, the most pleasing way of dividing a segment. The name itself seems to have been invented in Germany in the nineteenth century and the German critic Zeising claimed that it was the universal key to harmony. Only then did it begin to influence artists and architects on the conscious level, as did the Golden Rectangle, whose dimensions are similarly related.

We can base a golden rectangle on a square ABCD, dividing it into halves with the vertical EF and drawing an arc of radius EB with E as centre. If we lengthen the line DC to H where it crosses the arc and raise from H a vertical which crosses the extended line AB at G, we have the rectangle AGHD, which is golden. If we count AD as two units, DE and EC must each be 1 unit and, by Pythagoras, the diagonal EB must be the square root of 5 or 2·236. So must EH for they are both radii of the

same arc. The proportions of a golden rectangle are therefore
2:3·236 or 1:1·618. (Figure 10.)

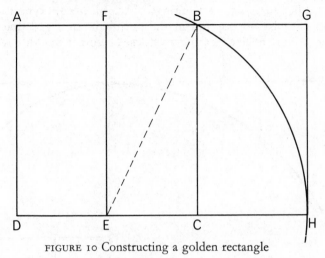

FIGURE 10 Constructing a golden rectangle

It is interesting to discover that this ratio is very close to that
between any two successive numbers of the Fibonacci series
and for practical purposes, we can use these as a sequence of
ready-made measurements for golden rectangles—3 units by 5,
5 by 8, 8 by 13, 13 by 21, and so on.

Golden rectangles are also interesting because we can derive
from them an infinity of smaller golden rectangles by repeatedly
cutting off a square based on the shorter side. By taking the
square ABCD from the rectangle AGHD, we are left with the
rectangle BGHC which is also golden. If we draw a line JK to
form a new square BGJK, the rectangle JKCH is golden too, as
is the rectangle KLMC when we remove the square JLMH, and
the rectangle KLPO when the square OPMC is taken away. If
we draw arcs with points C, K, L, and P as centres and with radii
equal to the successive squares, we shall form a spiral which comes
very close to being logarithmic. So once again, we are back with
the daisy and the nautilus shell. (Figure 11.)

We find what Pacioli called 'the divine proportion' of the
golden section in works of art created both before and after his
time. The façade of the Parthenon fits into a golden rectangle

and the section itself appears in the buildings of thirteenth-century Cistercian abbeys, in the sixteenth-century engravings of Albrecht Dürer, in the paintings of the nineteenth-century neo-Impressionist Seurat, and in the twentieth-century abstracts of Piet Mondrian. The fact that it was used before it was 'discovered'

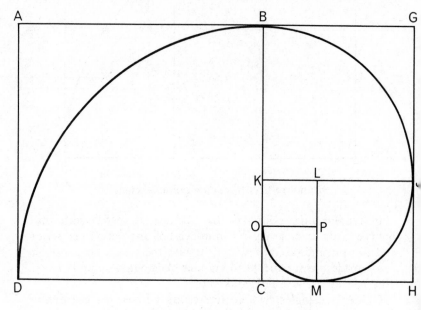

FIGURE 11 Successive golden rectangles

need not worry us, for Le Corbusier, 'the Picasso of modern architecture', has pointed out that rules of proportion are not always followed consciously. 'They may . . . be *implied* by the creative instinct of the artist, a manifestation of an intuitive harmony, as was almost certainly the case with Cézanne: Michelangelo being of a different nature, with a tendency to follow preconceived and deliberate, conscious designs.'[21]

Corbusier was intensely interested in the problem of proportion. As a student of 23, he asked himself: 'What is the rule that orders, that connects all things?' Thirty-six years later, he put the finishing touches to the Modulor, a system of basic measurements that architects could use in every kind of building.[22]

In itself, the idea of a module was not new. We find it in nature—in the hexagonal cells of a honeycomb, in the structure of plant tissues, in crystal formations. Men too have used modules from the earliest times. The Greeks and Romans took half the lower diameter of a column, divided it into thirty units and designed every other part of the building in terms of these. The scrolls of Ionic columns, for instance, always had a maximum radius equal to one quarter of the lower diameter.

Corbusier was more ambitious. He wanted to devise a module that was elegant, mathematical, and also human. In twentieth-century France, he was obliged to use a system of measures based on the metre which related to nothing but itself. Other civilizations had derived their measures from parts of the human body—the finger, palm, foot, and cubit (or forearm) and it seemed to him that these man-based measures led to a more human scale in architecture. When he ran his rule over the rooms of dwelling-houses from Turkey to Switzerland and even of the Petit Trianon and the *petits appartements* of the Faubourg Saint-Germain of pre-revolutionary France, he found them all about 2·20 metres high, roughly equivalent to the height of a man with upraised arms. How could he work out a mathematically-based module that would take the human body as its starting-point?

Between the wars, his talent and panache won him an international reputation as a leader of the *avant-garde*. His plans for projects that were never built often caused more of a stir than other people's realized schemes: for instance his idea of replacing run-down sections of Paris with vertical neighbourhoods set in parkland. He aroused violent opposition and his entry for the League of Nations building in Geneva was thrown out on the pettifogging ground that it was not drawn in Indian ink. Even so, he was continuously busy on schemes ranging from workers' houses at Pessai, near Bordeaux, France, to the Ministry of Education and Health in Rio de Janeiro. It was not until 1940, when Paris was occupied and his studio closed, that he was able to spend time on his early idea for a 'grid of proportions' which would serve every kind of tradesman working on the mass-produced buildings which he expected to spring up after the war.

He spelled out the problem to a young assistant, Hanning: 'Take a man-with-arm-upraised, 2·20m in height; put him inside two squares 1·10 by 1·10 metres each, superimposed on each other; put a third square astride these first two squares. The third square should give you a solution ... With this grid for use on the building-site, designed to fit the man placed within it, I am sure you will obtain a series of measures reconciling human stature (man-with-arm-upraised) and mathematics ...'[23]

Both Hanning and Mlle Elisa Maillard of the Musée de Cluny worked on the project and produced broadly similar solutions, each making use of the golden section. Taking a man of 1·75 metres in height, the man-with-arm-upraised became 216·4 centimetres and the width 108·2. The relationship between 175 and 108·2 was golden and therefore the basis for a Fibonacci-type series, which would give smaller or larger units, all related to the original height of 1.75 metres.

Corbusier thought the problem was now largely solved and took out a patent on his grid. His assistant, Soltan, was not convinced. It seemed to him that they had not a grid but a series of golden sections stretching from zero to infinity. It was not two-dimensional but linear. 'All right,' said Corbusier, 'let us call it henceforth a *rule* of proportions.' And there matters stood when he boarded the cargo-boat *Vernon S. Hood* on which he was to cross the Atlantic at the end of December 1945 as head of an architectural mission to America.[24]

The *Hood* had cabins only for the crew, and Corbusier, along with the other passengers, was assigned to a dormitory, hardly the place in which to work out the complicated problem now worrying him. Luckily, a fellow passenger persuaded a ship's officer to allow Corbusier the use of his cabin each day from 8 a.m. to noon and from 8 p.m. to midnight. Here, during a violent storm, he came on a solution which had the simplicity of genius. Using the nearest whole numbers, he took 108 as his basic unit for man-at-waist-level and doubled it to 216 for man-with-arm-upraised. He now had his two superimposed squares. Applying the golden section, he found that 216 divided into lengths of approximately 82 and 133 and by splitting down the 133 by further golden sections, he worked out a Fibonacci-type series descending to 82, 51, 31, 20, 11, 9, and 2. If required, it could be extended beyond 216 in a similar manner. This series, based on the double unit of 216, he called the Blue Series. He then constructed a second series on the single unit of 108. By applying the golden section upwards, he immediately arrived at 175, the head-height. Downwards, successive golden sections gave him 66, 41, 25, 16, 9, 7, and 2. He called this the Red Series and by putting the two scales side by side, he found he could draw in a perfect logarithmic spiral.[25] (Figure 12.)

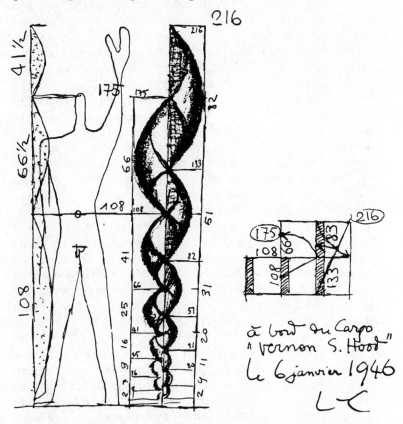

FIGURE 12 Corbusier's original sketches for the Modulor

Corbusier now had an infinitely flexible system. To form a grid, architects could combine red with red, blue with blue, red with blue or blue with red. And not just architects. His ideal was universal standardization on a human scale of cities, suburbs, homes, furniture, and every kind of manufactured article from a typewriter to a can of beans, from a newspaper to an aeroplane.

There was one snag. The world was still split between metric and imperial and the Modulor, as he called his system, did not translate easily into feet and inches. He and his colleagues were discussing the problem when one of them said: 'The values of the Modulor in its present form are determined by the body of a man 1·75m in height. But isn't that rather a French height? Have you ever noticed that in English detective novels, the good-looking men, such as the policemen, are always six feet tall?'[26]

They translated six feet into metric and found it came close to 183 centimetres. This head-height gave a waist-height of 113 by golden section, doubling to 226 for man-with-arm-upraised. The Red Series then became 10, 16, 27, 43, 70 centimetres and so on, with convenient near-equivalents of 4, 6½, 10½, 17 and 27½ inches. The Blue Scale fitted in equally well. The final drawing for the Modulor was produced by two of Corbusier's young assistants, one French, the other Uruguayan. 'There is not need to say more,' wrote Corbusier. 'It is enough to look.'[27] (Figure 13.)

But was it? The post-war plums to which Corbusier's reputation rightly entitled him were slow in coming. He drew up plans for rebuilding the ruined cities of Saint Dié and La-Pallice Rochelle in France but they were turned down by the authorities. For the United Nations building in New York, he was not even asked to submit a design but merely made a member of the jury that judged other people's. Later, he worked on major projects all over the world, from the National Museum of Western Art in Tokyo to the Carpenter Visual Art Centre at Harvard University; from the Olivetti building in Milan, Italy to Chandigarh, the new capital of the Punjab in India. In Britain, his influence can be seen in the 'new brutalism' of so many state schools and in the land-scaped tower-blocks of the Roehampton estate near London, an elegant example of a style of town-planning now out of favour.

As for the Modulor, it had its first full-scale trial in his *unité d'habitation* in Marseilles, a 'machine for living' on eighteen floors with every facility from shops to an open-air theatre serving the needs of its 1,800 inhabitants. Several other *unités* went up in France and in West Berlin but many architects disliked their garish colours, rough-cast concrete and starkness of design, castigating them as follies. They also pointed out that the Modulor was not, in practice, the 'rule that orders, that connects all things' which Corbusier had originally set out to find. Even so

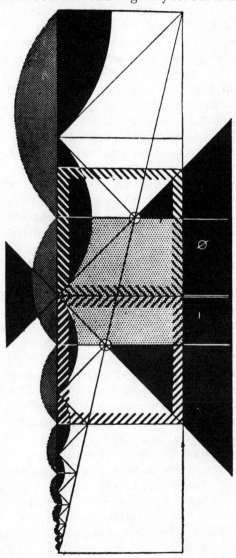

FIGURE 13 The final drawing of the Modulor

sympathetic a critic as Scholfield points out its limitations. The dimensions of individual apartments, roof structures, and some decorative sculpture followed the Modulor but little else. 'It integrates the whole structure in the sense that it links the parts to each other, but it fails to relate them to the whole.'[28]

Corbusier himself kept an open mind. He wrote: 'Assuming that the Modulor is the key to the "door of the miracle of numbers", if only in a very limited sphere, is that door merely one of a hundred or a thousand miraculous doors which may or do exist in that sphere, or have we, by sheer hazard, opened the one and only door that was waiting to be discovered? . . . I still reserve the right at any time to doubt the solutions furnished by the Modulor, keeping intact my freedom which must depend solely on my feelings rather than on my reason.'[29]

Beyond Modulor lies the wider question: can that element in a work of art which gives us aesthetic pleasure be predetermined, at least partly, by number? If so, do those proportions which we can express mathematically trigger some inborn response in the human psyche? Or are we conditioned to appreciate them?

Ernst Fischer, the Austrian writer and politician, took a strongly Marxist view. It was, he said, impossible to separate form from function. When man first wanted a vessel for water, he devised a primitive jug. Over the centuries, he gradually improved it until he had perfected an instrument for holding and pouring. He now had the ideal jug on which to model others. '*Form*,' said Fischer (his italics), '*is social experience solidified*.'[30]

Few of us would quarrel with that, so far as it goes, nor with his qualification that to some degree, form is modified by the materials used. It would be absurd to design a modern office-block of steel, glass, and concrete in the style of a rococo palace built in stone. But what of proportion and symmetry? Fischer insists that these too are determined by man's inherited experience and the materials he uses. 'A jerry-built, lop-sided house will last less well than one that follows certain laws of symmetry. Just as symmetry in crystals is the expression of an equilibrium of energy and hence of a saving of energy, so the symmetry of a house or other man-made object is also an expression of equilibrium.'[31] Then why do we find some schemes of proportion equally in favour at different periods of civilization when different materials are in use?

Because, says Fischer, man's tendency to be conservative leads to a time-lag between his discovery of new materials and his acceptance of an appropriate technique for working in them. Early man, for instance, built stone huts in a style which still preserved elements more suitable for huts of mud and straw. 'There is nothing surprising about this . . .: it is an extension of the tendency of all collectives to hold on to their hard-won social experience, to pass it down from generation to generation as a treasured inheritance.'[32]

The theory is ingenious but does it fit all the facts? Does it explain man's preference for the golden section which has persisted for thousands of years? The golden section is not essential to function and no one has ever suggested that buildings with other proportions are more likely to fall down. It pleases us equally whether it is applied to stone, brick, concrete, or paint on canvas. Experiments in which members of the public were asked to choose between a series of variously divided lines and rectangles of various proportions showed a clear preference for golden sections and golden rectangles. Attempts have been made to explain this preference physiologically, psychologically, and even psychoanalytically but they have been as unconvincing as Fischer's attempts to explain them in terms of function and material.

Corbusier himself believed that we favoured them because of a conditioning so profound that the preference had become a part of our nature. They governed 'a part of things which constitute the spectacle open to our eyes—the ramifications of a leaf, the structure of a giant tree or a shrub, the bone structure of a giraffe or a man—things which have been our daily bread, or our exceptional visual experience, for millions of years. Things which constitute our environment . . .'[33] It seems then that we inherit a sense of these proportions and feel an aesthetic pleasure when we find them again in the natural world or re-create them in work of our own. 'Numbers,' says Corbusier, 'lend dignity to the houses of men. They make a temple out of an ordinary dwelling.'[34]

The British art critic Herbert Read takes a similar view but puts it rather differently. Beauty, he maintains, does not exist in nature but is created by man. 'It is a man's sensuous apprehension of the godlike; the result of his assumption of a *constructive* function. In nature (which is what is given to man) the artist finds measure, which is a reality he has learned to express in number.' He can do this in one of two ways, both of them intuitive. He can express numerical relationships directly, a technique which we can trace in a tradition stretching from ancient architecture to the paintings of Seurat and Cézanne and the sculpture of Maillot and Rodin. The alternative, which is equally valid, starts from an idea. 'It does not reject measure (any more than the first mentioned point of departure ignored idea) but it prefers to follow the path indicated by organic or biological evolution. Nature has been selective in its manipulation of geometrical data; growth is not amorphous, but restricted by a limited number of physical laws. These might be called environmental laws; the laws which determine the inter-relations of chemical substances. The egg is not an ordinary shape; it is determined, as we say, by physical laws.'[35]

To point out the difference between the two approaches, Read elaborates the example of the egg in full technical detail. The shape of its shell can be expressed mathematically and if Pn is the normal component of external pressure at a point where r and r^1 are the radii of the curvature, T the tension of the envelope, and P the internal fluid pressure, we find that

$$Pn + T(1/r + 1/r^1) = P.$$

Without actually making any calculations, the artist could take these proportions as found in a particular egg and use them for drawing something quite different, say a television set or an abstract form. Alternatively, he could mentally give T and P different values and draw an egg-like object according to the formula. 'It is really a choice between applying the values of a particular formula, or varying the values of a general formula.'[36]

According to this view, which would probably be shared by most Westerners, proportion in art is the result of a continuing interaction between man and nature. We cannot produce art by expressing mathematical relationships mechanically, for they are only one ingredient among many, but the fact that we can measure them and find them not only in the natural world (the daisy and the nautilus shell) but in the abstract concepts of mathematicians and in acknowledged works of art which have come down to us over the millennia, suggests that they are not just functional.

Corbusier asked: 'May a visual work, intended for the delectation of the spirit, making use of form, of broken-up surfaces, of holes and protuberances, in short: of measurable elements arranged in harmony or opposition (I am speaking of architecture and painting)—may such a work owe its being, in part to geometry and mathematical relationships?'

He went on to answer his own question. 'The reply is, of course, in the affirmative; that is in the nature of things.'[37]

9

Is God a Number?

IN April 1976, readers of *The Times* were treated to a fascinating titbit of literary detective work. Bishop Mark Hodson wrote:

> 'If you look up Psalm 46 in the Authorized Version of the Bible and count 46 words from the beginning of the psalm, you will find that you have arrived at the word "Shake". Now, discounting the word "Selah" [a musical direction], count 46 words from the end of the psalm and the word then revealed is "spear".'

Is there any possibility that the words were placed in these positions by pure chance? It seems unlikely. For the translators finished their work in 1610 when Shakespeare was 46 years of age.

> 'I should say that they must have known him not only as a great playwright and poet (1610 was the year the sonnets were first published), but also as a friend. They knew he would enjoy with an amused pleasure this charming little tribute.'[1]

If this sounds like one of those dotty schemes for predicting the date of your grand-daughter's wedding by adding together your post-code, your telephone number, and the square of your body-weight divided by π, it is worth bearing in mind that cryptograms were common in mediaeval and Renaissance literature. Shakespeare himself used numbers to convey hidden meanings in his poems and so did Spenser, Milton, and Donne among many others. They drew on a tradition both scholarly and popular that the entire cosmos was ruled by number and that any individual numbers mentioned, however casually, must give added meaning to the text. Sometimes, the structure of a poem was built on numerological principles from the number of feet in a line to the way in which stanzas and even books were arranged.

Belief in the power of numbers can be traced back at least as far as Pythagoras who lived in the second part of the sixth century B.C. He and his followers held that things *were* numbers or at least resembled them. They were familiar with triangular, square, and oblong numbers and knew that the square on the hypotenuse of a right-angled triangle equalled the sum of the squares on the other two sides. They probably discovered the 3:4:5 triangle from the work of Egyptian mathematicians and found for themselves the 5:12:13 and 8:15:17 triangles which followed the same rule. It was their discovery of the link between music and the first four whole numbers (see chapter 8) that led them to believe that number ruled the world, and by this they always meant whole numbers.

To understand their awe at the discovery, we must remember that they were not just mathematicians and scientists but members of a religious cult, pursuing their search for truth by the light of inner experience and intuition. They believed in reincarnation and the transmigration of souls. At their school at Croton in southern Italy, they seem to have worshipped Apollo and the Muses, using rituals derived from the mystery religions of Thrace and observing a strict rule that covered diet, clothing, and every form of conduct. This, they believed, was the only way in which the soul could achieve its destiny, which was union with the divine.

Number was a Jacob's ladder which the soul must climb on its journey and it was their job to find the individual numbers ruling every aspect of life. The cosmos was the One, which was conceived of as being neither odd nor even. It was also reason, whereas 2 was dissension ('it takes two to make an argument'). Four was justice because it was doubly equal, being $2 + 2$ and also 2×2. Seven was opportunity, for man developed in multiples of seven years, and so on.

Most of all, though, the Pythagoreans revered the *tetraktys*, a triangular figure consisting of rows of one, two, three, and four dots (Figure 1). It was as important to them as the Cross to Christians, for it symbolized the four elements—earth, air, fire, and water. The first four numbers also symbolized the harmony of the spheres and added up to ten, which was unity of a higher order. It is said that initiates were required to swear a secret oath by the *tetraktys* when they began their three years of silence as novices, and they even prayed to it.

FIGURE I The tetraktys, symbolizing to the Pythagoreans the four elements.

'Bless us, divine number, thou who generatest gods and men! O holy, holy *tetraktys*, thou that containest the root and source of the eternally flowing creation! For the divine number begins with the profound, pure unity until it comes to the holy four; then it begets the mother of all, the all-comprising, all-bounding, the first-born, the never-swerving, the never-tiring holy ten, the keyholder of all.'[2]

All through the fifth and fourth centuries B.C., the Phythagoreans pursued their researches, many of them fruitful and valid. In music, they worked out the intervals of the diatonic scale and defined both the twelve-tone chromatic scale and later the enharmonic scale, which unlike the modern piano distinguishes between G sharp and A flat. In astronomy, a fifth-century Pythagorean, possibly Philolaus of Croton, anticipated Copernicus by asserting that the earth was not the centre of the universe, proposing instead that it revolved round a central fire together with the moon and fixed stars.

Ultimately, Pythagoreanism foundered on the very rock on which it had been built. Its belief in the all-embracing power of whole numbers, revealed by simple musical harmonies and confirmed by the $3:4:5$ triangle, became too rigid. No exceptions were admitted and the Pythagoreans continued to apply it with religious fervour even when it did not fit the facts. This led to jiggery-pokery and even suppression of the truth.

The planets had always had a special significance for the ancients and it was usually reckoned that there were seven of them: sun, moon, and the five planets then known. This was not good enough for the Pythagoreans. The planets were so important in the scheme of things that they must correspond to the sacred *tetraktys*. To bring the number up to ten, they added the complete sphere of fixed stars, the earth itself and a wholly imaginary counter-earth around which earth was supposed to circle. We never saw it, they said, because the parts of earth inhabited by man always faced in the opposite direction.

If faith can move mountains, it might be expected to create heavenly bodies but soon the Pythagoreans were put to an even fiercer test. Ironically it came with further investigation of the right-angled triangle by which the name Pythagoras is now best remembered. They made an important mathematical discovery but in so doing blasted the foundations of their own system. For when they applied the 'Pythagoras' formula to right-angled triangles formed by bisecting a square diagonally, they found that the hypotenuses could never be expressed in terms of the all-sacred whole numbers. If the two sides of each triangle were each equal to one unit, the hypotenuse must be equal to $\sqrt{1^2 + 1^2} = \sqrt{2}$, and there was no way of expressing $\sqrt{2}$ in terms of whole numbers. This may not seem important to us because we are familiar with irrational numbers. For the Pythagoreans, who based their entire faith on whole numbers, it was anathema. They had opened Pandora's box and let fly a type of number that far from being whole was not even expressible.

They dubbed these irrationals *alogon* or unutterable. It was said that their discoverers drowned in a shipwreck and that Hippaeus, who insisted on studying them, was expelled from the community. Though some of the Pythagoreans had the courage to accept their existence and even found a method of working out an approximate value for $\sqrt{2}$, the irrationals dealt the school a blow from which it never recovered and the serious study of mathematics gradually declined.

As a religion, though, Pythagoreanism was still alive and kicking. It survived in the form of a mystery cult, one of many that flourished throughout the Roman Empire, and some of its elements can still be traced in the beliefs of present-day theosophists and occultists. Its number theory was even more influential for it played a major role in Plato's cosmology. Eventually,

Pythagoreanism and Platonism became so intermingled that it is often impossible to say whether a given philosopher owed more to one or to the other.

As we know from the *Timaeus*, Plato firmly believed in the harmony of the spheres expressed in terms of number. His world model is somewhat obscure and scholars have never agreed on all the details but the broad principles are clear and we can construct a reasonably coherent picture. He accepted the view of Empedocles that there were four elements and when the Demiurge (or prime mover) created the universe, he used earth to give it solidity and fire to make it visible. These two had to be joined together not by one other element, for that would have implied a flat universe, but by two. So air and water were set between earth and fire.

Now the universe was as near perfect as possible. Its overall shape must therefore be spherical and the elements forming it joined 'in the fairest manner'. The ideal proportions were expressed numerically.

> 'For whenever the middle term of any three numbers, cubic or square, is such that (a) as the first term is to it, so it is to the last term,—and again (b) conversely, as the last term is to the middle, so is the middle to the first,—then (c) the middle term becomes in turn the first and the last, while the first and the last become in turn middle terms, and the necessary consequence will be that all the terms are interchangeable, and being interchangeable, they all form a unity.'[3] (My sub-letters.)

As an example of the kind of relationship he has in mind, we can take the sequence 2, 4, 8. The arrangement (a) then gives us 2:4:4:8; (b) gives 8:4:4:2; and (c) 4:8:2:4.

This then is Plato's model of the body of the universe. Its soul started as a strip of Being divided into seven portions, the first being unity and the rest equal to powers of 2 and 3 alternately, giving a succession of 1, 2, 3, 4, 9, 8, and 27 which are traditionally arranged in the capital form of the Greek letter *lambda* (Figure 2). Between these were inserted further portions of Being representing arithmetical and harmonic means (Plato does not say how) so that the whole series corresponded to a musical scale covering four octaves and a major sixth in proportions as abstruse as 243/128. The strip was divided vertically and the two resulting strips crossed at right angles, bent back on themselves and joined to form two bands, like the circles formed by the equator and o° longitude on the globe. So the world soul neatly fitted the spherical world body (Figure 3).

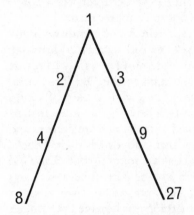

FIGURE 2 The Platonic *lambda*

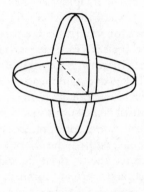

FIGURE 3 The world Soul

The vertical band was divided into a series of motions related to each other by the ratios of the *lambda* and these became the orbits of the seven planets which were then created 'for the determining and preserving of the numbers of Time'. The planets moved at various speeds but 'the ratios of one to another are those of natural integers'.[4] As far as possible, therefore, Plato's universe is created in terms of whole numbers.

The scheme is complicated and speculative and if it seems a little dry in summary, the full account is among the most moving and poetic of all creation myths. It also has the merit of fitting what were then accepted facts about the material universe, even though it sometimes seems strained with its outlandish intervals between the planetary orbits. Plato may have been inspired by the vision of a divine harmony but he never let it drive out the results of observations, meagre though these were.

He did, however, let loose a flood of wild number theories. The less cautious of his followers bundled up his ideas with those of Pythagoras and drew whatever conclusions suited their own tastes. Some took the view that every number must have a meaning beyond itself. Others believed that they could themselves rule the world by manipulating number. Magicians and occultists of every hue were swept up in a number craze that was to last for the next two thousand years. The imagination of serious thinkers was also fired, for number, as we shall see in the next chapter, has a power over the imagination that is still not fully appreciated.

Jews of the Hellenistic world especially found number a useful tool for validating the scriptures. They had scornfully dismissed the ancient Greek legends as man-made nonsense, only to realize that there were stories of a similar kind in the Old Testament. They were convinced that these stories were a part of God's revelation but proving it to non-believers was another matter. Luckily, the scriptures were packed with number references and these were interpreted in ways that supported their beliefs. Philo justified the story of Genesis by pointing out that God created the world in six days. This *must* be correct because 6 was the first perfect number, being the sum of its own divisers $= 1 + 2 + 3$ (see chapter 7). In numerology, 6 became the number of Creation and no one worried that the philosophizing of pagans was being used to reinforce the claim that Jahweh was the one true God.

Many of the Christian Fathers continued in the same tradition. They believed that the Bible summed up the whole of man's knowledge about God and the universe and any numbers mentioned must therefore have a special significance, which it was their duty to find. For was it not said, 'Thou hast ordered all things in measure and number and weight'? (Wisdom of Solomon, 11.20.)

St Augustine was an adept in the numerological interpretation of the Bible. There was, for instance, the incident in John 21.11 where the Apostles landed 153 fish. To most of us, 153 might seem a pretty hard number to explain but it held little difficulty for Augustine. The clue lay in the number of disciples hauling in the nets. There were 7 of them. We know that there are seven gifts of the Holy Ghost and these enable men to obey the ten commandments. The disciples must therefore be saints. Moreover, $10 + 7 = 17$ and if we add together the numbers 1 to 17, we get a total of 153. The hidden meaning of the text, therefore, is that 153 saints will rise from the dead on the last day.[5]

Another of Augustine's expositions comes from his *City of God* which is symbolized in the Bible by the Ark.

'For the dimensions of the length and breadth of the ark signify man's body, in which the Saviour was prophesied to come, and did so; for the length of man's body from head to foot is six times his breadth from side to side, and ten times his thickness measuring perpendicularly from back to front. Lay a man prone and measure him, and you shall find his length from head to foot to contain his breadth from side to side six times, and his height from the earth whereon he lies, ten times; whereupon the ark was made 300 cubits long, 50 broad and 30 deep.'[6]

Scholars did not stop at explaining the number systems used by others. They carried on the tradition in their own writings. As Christopher Butler points out, the *City of God* itself has a pattern that can be explained in terms of number symbolism:

'It has 22 sections (perhaps symbolizing completeness since 22 is the number of letters in the Hebrew alphabet). These comprise two groups of five devoted to refutation (the ten negative precepts of the Law), and three groups of four dealing with positive teachings (symbolizing the twelve apostles and the four Gospels.'[7]

Such an interpretation may seem as fanciful as some of those propounded by Augustine himself but there is good reason for thinking that he practised what he preached. It is now recognized that scores of mediaeval and Renaissance works were steeped in number symbolism and the only question still open to argument is not 'whether' but 'how much'?

We also find numerology strongly influencing early Christian art, especially that of the Byzantines who inherited a potent strain of Neopythagoreanism through the mathematician Nicomachus of Gerasa who lived in the first century A.D. The number to look for is 6, the perfect number of Creation, and we find it in, say, a six-sided bowl or a group of six sheep in a mosaic. It also occurs in the proportions of a mosaic, commonly 3:2 or 6:1. Father Gervas Mathew wrote:

> 'Six also represented the process of the Incarnation; multiplied by ten, it became the symbol of Christ. The harmony of six consist in the harmony of its components; ... Byzantine "surface-aesthetic" was inevitably derived from arithmetic since all harmonies in form or colour were the echoes of an incorporeal music, the harmonies of Pure Number.'[8]

The appreciation of these harmonies was not confined to the central Judaeo-Christian tradition. In various ways, the ideas of Pythagoras and Plato were taken up by almost every mystery religion and especially by the Gnostic sects which reached the height of their influence in the second century A.D. They mingled Eastern, Jewish, Hellenistic, and Christian ideas in various combinations and believed that only Gnosis (true knowledge) could save man from the evils of the world. They were usually denounced as heretics by orthodox Christians and it was all the more necessary, therefore, for them to justify their beliefs by reference to the New Testament. This was especially true of their cosmologies which were sometimes even more complex than Plato's.

One of the most interesting was that of the Valentinians, followers of Valentinus, an Alexandrian who hoped to become bishop of Rome and left the Church when he failed to do so. They believed that the creator-God of the Old Testament, whom they called the Demiurge, was a fatherless abortion of Sophia, youngest of thirty Aeons who ruled the Pleroma or spiritual sphere. The Aeons were arranged in three groups of 4, 5, and 6 male-female pairs or syzygies, making $8 + 10 + 12 = 30$ in all. The Demiurge was produced when Sophia deserted her Aeon-partner Willed and lusted after Depth, the original, eternal Aeon who could be comprehended only by his son Mind. The material from which the world was created was born of Sophia's ignorance, fear, and grief and was essentially evil until redeemed by Christ.[9]

This is the simpler of at least two versions of the Creation attributed to the Valentinians, though all have in common the thirty Aeons. Some writers have found parallels with Persian sects and it has also been suggested that the 8 and the 12 corresponded to the primary and secondary gods of the Egyptians. There seems also to be a Pythagorean element, for the first four pairs and the second ten Aeons can be seen to echo the *tetraktys*. Perhaps the most likely explanation is that all these influences played a part, and that knowingly or otherwise, the Valentinians were combining a number of systems that were originally separate.[10]

However that may be, they had to prove the truth of their scheme by reference to the Bible. For starters, they pointed out that the Saviour embarked on his ministry when he was thirty years of age (Luke 3.23) and in the parable of the labourers of the vineyard, men were hired about the first, third, sixth, ninth and eleventh hour (Matthew 20.1–16), giving $1+3+6+9+11=30$. So the total number of Aeons was confirmed twice over.

The fact that the first groups contained eight and ten Aeons posed a difficulty but the Valentinians found some consolation in the Greek form of the name of Jesus whose first two letters are *iota* and *eta*, then widely regarded as standing for 10 and 8 respectively. This, of course, was the wrong way round but it was the best they could find and if they had any doubts, these were swiftly resolved by the numerous proofs for the correctness of twelve, the number of the third group. For Jesus had first preached in the Temple at the age of twelve and later had twelve disciples. Moreover, the last of these, Judas Iscariot, betrayed him, just as the last of the Aeons, Sophia, betrayed her partner Willed. If that weren't proof enough, the Valentinians pointed out that Jesus preached for twelve months before his crucifixion and that the woman with an issue of blood suffered for twelve years before being healed. These were both in their ways comparable with the agony suffered by Sophia when she found that she was unable to attain knowledge of Depth, the forefather.[11]

During the mediaeval period, new influences widened and deepened the numerological tradition. One was the body of writings first thought to have been the work of a priest of the Egyptian god Thoth, known in Greek as Hermes Trismegistos or 'thrice-greatest Hermes'. He was supposed to have revealed the secrets of Chaldean oracles long pre-dating Plato who was said to have drawn on these for his *Timaeus*. We now know that the Hermetic writings were the work of Gnostic sects of the first,

second, and third centuries A.D. and themselves drew on Plato's
Timaeus and other sources.

Equally influential was the Jewish mystical tradition or
Kabbalah. One of the ways in which Kabbalists interpreted
scripture was to give each Hebrew letter a numerical value, a
method known as *gematria*. In Genesis 49.10 we find the passage:
'The sceptre shall not depart from Judah, nor a lawgiver from
between his feet, until Shiloh come; and unto him shall the
gathering of the people be.' In Hebrew, 'until Shiloh come' is
IBA ShILH (*Yavo Shilo*) and if we add together the numerical
values of the letters, we get a total of 358. Now the Hebrew for
Messiah is *MshICh* (*Meshiach*) and these letters also total 358. The
Kabbalists therefore took the phrase in Genesis as a prophecy of
the coming of the Messiah. Some Christians also practised *gematria*
and since the Hebrew name of the Serpent of Moses, *NChSh*
(*Nachash*), also totals 358, they linked it with the Crucifixion. In
mediaeval iconography, we sometimes find the symbol of a
serpent wound round the cross.[12]

Kabbalists also used *gematria* to expound the *tetragrammaton* or
four-lettered name of God. It was written *IHVH* which became
Jahweh or Jehovah in English but was originally thought to be
so holy as to be literally unspeakable. It was remembered only
because the elders were allowed to pass it on to their disciples
once in every seven years. Yet, according to the Kabbalists, it was
itself a substitute for a much longer form of the holy name, the
Shem ha-meforash which had seventy-two syllables and 216
letters.

'The source of the *Shem ha-meforash*, according to tradition,
are verses 19–21 of Exodus 14, each of the three verses containing seventy-two Hebrew letters. The letters of verse 19
were written down in separated form and in correct order;
the letters of verse 20, also in separated form, were written
down in reverse order, and the letters of verse 21 were
written down in correct order. Reading from above down,
one obtains seventy-two three-letter names, all of which
combine to make one. To these three-lettered names were
then added either AL or IH to form the names of the seventy-
two angels of Jacob's ladder.'[13]

Happily, if the individual letters of *IHVH* are arranged in the
form of a *tetraktys*, given their numerical equivalents and added
together line by line, the total comes to 72. 'Therefore, the correct
pronunciation of the four-letter name of God is thought of as
being just as effective as the correct pronunciation of the *Shem
ha-meforash* because gematria shows that the latter is contained
within it.'[14] Kabbalists used the *tetragrammaton* and the other
seventy-two names both as objects of contemplation and as a
means of invoking God and his angels to work magic on their
behalf.

All these traditions came together in the Renaissance. As
Dr Francis Yates has shown, they were drawn on by a series of
thinkers such as Cornelius Agrippa, Ficino, Pico della Mirandola,
and Giordano Bruno, but they were not confined to a few
esoterics.[15] They were the high fashion of Renaissance philosophy.
At one level or another, numerology was accepted by everyone
from the most learned academics to the least knowledgeable
laymen.

There were, of course, discrepancies because different systems
gave different values to the same numbers. One was sometimes
thought of as unity or perfection and 2 as divisive because it
marred this perfection, bringing about strife. Other numerologists regarded 1 as defective, needing another 1 to complete it
in the fullness of 2. In many systems, 2 is feminine and 3 masculine.
A combination of 2 and 3 will therefore stand for sexual union or
marriage. But is it to be 2 × 3 or 2 + 3? Some thought 6 should be
the conjugal number, others 5. On the whole, the 5s had it as
Sir Thomas Browne made clear in *The Garden of Cyrus or The
Quincuncial, Lozenge or Net-work Plantations of the Ancients Artifici-
ally, Naturally and Mystically Considered* (1658). Appropriately, he
leaves discussion of the quincunx, five dots in the arrangement
found on a die, until chapter 5.

'He that forgets not how Antiquity named this the Conjugall
or wedding number, and made it the Embleme of the most
remarkable conjunction will . . . afford no improbable
reason why Plato admitted his Nuptiall guests by fives, in
the kindred of the married couple.

'And though a sharper mystery might be implied in the

Number of the five wise and foolish Virgins, which were to meet the Bridegroom, yet was the same agreeable to the Conjugall number, which ancient Numerists made out by two and three, the first parity and imparity, the active and passive digits, the materiall and formall principles in generative Societies. And not discordant even from the customes of the *Romans*, who admitted but five Torches in their Nuptiall solemnities. Whether there were any mystery or not implied, the most generative animals were created on this day, and had accordingly the largest benediction: And under a Quintuple consideration, wanton Antiquity considered the Circumstances of generation, while by this number of five they naturally divided the Nectar of the fifth planet.

'The same number in the Hebrew mysteries and Cabalistical accounts was the character of Generation; declared by the letter *He*, the fifth in their alphabet . . .'[16]

Renaissance poets were just as conscious of number symbolism and drew on it for both the content and form of their work. One of the nicest examples comes in Donne's *The Primrose* which describes how he goes out walking to find a true love among the flowers, which have different numbers of petals. As a man, his own number is 3 and he hopes to discover a flower with the correct conjugal number of 5, so implying union with a true woman, whose number would be 2. Until he comes on a primrose, which has five petals, he is perplexed, for all the other flowers have either six petals or four.

> 'Yet know I not, which flower
>   I wish: a six, or four;
> For should my true-love less than woman be,
> She were scarce anything: and then, should she
> Be more than woman, she would get above
>   All thought of sex, and think to move
>   My heart to study her, and not to love;
> Both these were monsters; since there must reside
> Falsehood in woman, I could more abide,
> She were by art, than nature falsified.
>   Live primrose then, and thrive
>   With they true number five . . .' (My spelling.)

The symbolism of 5 is further stressed by the poem's structure, alternating couplets and triplets.

The detection of numerological patterns in the structure of Renaissance poetry has become an academic growth industry and though we need not agree with all the discoveries claimed, it is impossible to deny that there are far more than was realized until a few years ago. Christopher Butler has ably reviewed some of the most interesting, including those of the Norwegian scholar Maren-Sofie Røstvig, who found it significant that the Lady in Milton's *Comus* should take exactly sixty lines to assert her chastity when she heard the goings-on of midnight wassailers in a 'wild wood'. Since there were Ten Commandments and 6 was a perfect number, 60 must symbolize 'the perfect fulfilment of the law'.[17]

Miss Røstvig is a serious and capable scholar but personally I am more convinced by the findings of her American colleague Professor Kent Hieatt who analysed Spenser's *Epithalamion*. The poet wrote it to celebrate his own wedding on 11 June which was St Barnabas's Day or 'Barnaby Bright'. In the sixteenth century, astrology had an even greater following than numerology and in a largely agricultural society without benefit of electric light, people were more conscious of the relative lengths of day and night. Bearing in mind that the Old Style Calendar was still in use, we find that Barnaby Bright was the longest day of the year and under the zodiacal sign of Cancer the Crab. Spenser shows his awareness of these facts in stanza 15:

> 'This day the sun is in his chiefest height,
> With Barnaby the bright,
> From whence declining daily by degrees,
> He somewhat loseth of his heat and light,
> When once the Crab behind his back he sees.' (My spelling.)

The relative lengths of day and night throughout the year could easily be found in contemporary almanacks and for St Barnabas's Day, they were reckoned as $16\frac{1}{4}$ hours and $7\frac{3}{4}$ hours respectively. Given Spenser's evident interest in the subject and the numerological tradition in which he worked, we need not be surprised to find these proportions reflected in the structure of his poem.

This does in fact happen. *Epithalamion* has twenty-four divisions representing the twenty-four hours of the day. Twenty-three of them are full-length stanzas usually of eighteen lines and the last a short *envoi* of seven lines. Each stanza ends with a two-line refrain, the second being, with slight variations:

> 'That all the woods shall answer and their echo ring.' (My spelling.)

Significantly, the refrain always takes a positive form in the first sixteen stanzas. In the last seven, it is invariably negative, for instance:

> 'Ne will the woods now answer, nor your echo ring.' (My spelling.)

In other words, night and silence have descended on the woods at approximately the time we should expect from the information given by the almanacks.

Nor does the case rest merely on an approximation. Line 5 of stanza 17 begins:

> 'The night is come . . .'

After $16\frac{1}{4}$ stanzas measured to the exact half line, the day of $16\frac{1}{4}$ hours is over. The other $7\frac{3}{4}$ divisions are concerned with the $7\frac{3}{4}$ hours of night.

The correspondence between the structure of the poem and the period of time spanned by the action is only one example of many pointed out by Professor Hieatt. As Alastair Fowler has also shown, Spenser's numerology is not a whimsical ornament but integral.

> 'The analogy between the repetition of the seasons and the perpetuation of life through human generation is a part of the content; so that the numerology which imitates the year's cycle makes a dynamic contribution to the unfolding of the meaning . . . numerological form *participates* in the action of *Epithalamion*.'[18]

Fowler himself wrote a 300-page analysis of Spenser's *Fairie Queene* on numerological lines. He found it 'an astonishingly complex web of interlocking numerical patterns of many different kinds.' He went on:

> 'We find numerological significance in line-, stanza-, canto- and book-totals; in the location of these units; and even in the number of characters mentioned in each episode. Pythagorean number symbolism, astronomical symbolism based on orbital period figures and on Ptolemaic star catalogue totals, medieval theological number symbolism; all these strands and more besides, are worked together into what—in this respect at least—must be one of the most intricate poetic textures ever devised.'[19]

Fowler, who also collaborated with Christopher Butler on a numerological study of Shakespeare's *Venus and Adonis*, makes the point that the numerological aspects of *Fairie Queene* have been overlooked for generations, even though the text is riddled with clues: names such as Una and Duessa, strings of mathematical terms and even explicit numerological passages. We can blame our failure to see what was once so obvious on the Copernican revolution. The view of the Polish astronomer Nicolaus Copernicus (1473–1543) that the sun, not the earth, was the centre of the planetary system shattered beliefs that had persisted for some two thousand years. Though number and the music of the spheres were still important to Milton in the second half of the seventeenth century, *a priori* theories about the cosmos gave way to experimental science. This brought such rich material rewards over the next few hundred years that men forgot that alternative ways of looking at the universe were possible. They not only rejected the ideas of Plato and Pythagoras themselves: they were blind too to the fact that so much of their predecessors' work, both in science and the arts, had been built on them. Only now are such scholars as Hieatt, Fowler, Butler, and Dr Frances Yates rediscovering the full extent of the role that numerology and other such systems once played in intellectual life.

As many of the best minds abandoned numerology, it was gradually debased into the pseudo-intellectual parlour game satyrized by Tolstoy in *War and Peace*. He quotes one system of many for giving numerical equivalents to letters of the alphabet:

a	b	c	d	e	f	g	h	i	j	k	l	m
1	2	3	4	5	6	7	8	9	9	10	20	30

n	o	p	q	r	s	t	u	v	w	x	y	z
40	50	60	70	80	90	100	110	120	130	140	150	160

Bill Smith could thus find his lucky number by adding together
$2+9+20+20+90+30+9+100+8 = 288$, which by totalling the
digits becomes 18 and then 9. His full name William Smith,
however, would give $130+9+20+20+9+1+30+90+30+9+$
$100+8 = 456 = 15 = 6$. So Mr Smith can take his choice between
either 9 or 6, whichever proves the most convenient.

Fiddling of this kind was used to make Napoleon the anti-
Christ. L'Empereur Napoleon gave $20+5+30+60+5+80+5$
$110+80+40+1+60+50+20+5+50+40 = 661$, which seemed
innocent enough until someone remembered that an 'e' had been
dropped from the definite article. Add an extra 5 for this and you
get 666, the Number of the Beast. Q.E.D.

Innumerable number patterns can be found in history if we play
around long enough. In 1944, someone discovered an interesting
coincidence in the careers of the main national leaders:

Name	Year born		Age		When took office		Years in office		Total
Churchill	1874	+	70	+	1940	+	4	=	3888
Hitler	1889	+	55	+	1933	+	11	=	3888
Mussolini	1883	+	61	+	1921	+	23	=	3888
Roosevelt	1882	+	62	+	1933	+	11	=	3888
Stalin	1897	+	65	+	1924	+	20	=	3888

Now, 3888 was twice 1944, the year in question, Obviously, the
logic of history must be pointing to some lesson and wishful
thinking concluded that this was a prediction that the war would
come to an end in 1944. If the year of the war's end could be
found by dividing the magic 3888 by 2, we could presumably
discover the month, day, and hour by halving again, giving us
972. The war would therefore end at 2 a.m. on 7 September 1944.
It didn't.

Not all recent numerology has been so trivial. Serious occultists such as the followers of Madame Blavatsky and latter-day Kabbalists have taken the study as seriously as their great predecessors. Little is heard of them because they meet in inconspicuous groups, rarely advertising, and making themselves known only to those who have shown a respectful interest in the subject. No one who investigates it could doubt the thoroughness of their work.

In 1917, Mr F. B. Bond, a British architect, and the Reverend Dr Thomas Lea, then vicar of St Austell in Cornwall, England published the findings of 'a preliminary investigation' into 'a gematria in the Greek text of the New Testament'. They applied a method of substituting numbers for the letters of the Greek alphabet which they traced back to the third century A.D. and beyond that to a fifth-century B.C. source in the 'Syro-Phoenician centre east of the Mediterranean'.

As one example of their findings, let us take the name 'Cephas', which comes from John 1.42 where Jesus tells Simon: 'Thou shalt be called Cephas, which is by interpretation, A stone.' Numerologically, Cephas ($KH\Phi A\Sigma$) is equivalent to 729, which is the cube of 9 and is seen by the authors as the 'perfect stone on which Christ builds his church'. They point out that the Greek words for 'perfected work', 'temple of the church', and 'ministers of the church' also have numerical equivalents of 729. If we construct a cube of $9 \times 9 \times 9$ and view it cornerwise from above, we find that it reveals 243 facets of its component cubes out of a total of 486 facets if we could see all sides. (Figure 4, where the computations for the Greek words are also given.) The authors write:

'The 486 is $\Pi ETPA$ ($80+5+300+100+1$), so that PETRA, the Rock, is the surface of the cube whose solidity is CEPHAS. And $\Pi ETPA$ $\sigma\mu\gamma$ [$\sigma\mu\gamma$ is the Greek-letter equivalent of 243] or the Rock of 243 stones $= 729 = KH\Phi A\Sigma$.

'But it will be observed that the 243 facets visible belong to 217 separate cubes of which the central one shews Three facets and three rows of Eight cubes each (all marked in the Schema) shew Two facets. The central Cube represents the Chief of the Seven corner-stones and the triple Divine

Potency, whilst the Three Ogdoads of 8,8,8, all having a duality of aspect, stand for the 888 of $IH\Sigma OY\Sigma$—Jesus, Who united in His nature God and Man. Each arm of the Ogdoads encloses 8×8 stones = 3 times 64—$A\Lambda H\theta EIA$—a triple Truth, and the whole is $192 = MAPIAM$ the name of Mary, the Mother of Jesus, here symbolic of His Church. On the further side of the cube, invisible, are 169 more of the lesser cubes, and this is the number of the name of Christ. $\dot{O}$ $AMHN$ —the Amen—meaning Truth or Verity, and these 169 cubes show again 3×64 faces or a triple Aletheia which is Mariam.

'These $217 + 169$ cubes completely surround and enclose the Cube of $7 \times 7 \times 6 = EKK\Lambda H\Sigma IA$—the Church. Hence the Sevenfold Symbolism of the Ecclesia, otherwise evident in the Seven Corners of the Cube and the Seven Stars . . .'[20]

All this is just a tiny part of a numerological dissertation on such topics as 'The First Mystery', 'The Cabala of the Fish', 'The Names, Epithets and Types of Christ as Multiples of 37', 'The Cabala of the Cosmos', and 'The Cube of Light'. I have quoted it in full to show how an elaborate *gematria* can be taken up in the twentieth century not just by Kabbalists and Theosophists but by orthodox Christians, though they are, of course, far from typical.

It is easy for non-believers to poke fun but we must be careful. Even the most naïve methods of substituting numbers for letters and so divining a person's destiny from his name can be valid for those who believe in predestination and reincarnation (and if you have argued against such beliefs with an adept, you will know how hard they are to refute). Moreover, there is strong evidence that some people have a special power, call it psychic, supernatural, intuitive, telepathic, or what you will. These people are often stimulated by sets of symbols. While one draws insight from the *I Ching* or the Tarot pack of cards, another finds help in astrology. Numbers too are symbols and it may be that these act on some people in a similar way.

Nor has popular numerology disappeared entirely. The power of some numbers reverberates down the corridors of history and still influences our language and sometimes, too, our behaviour. Public opinion surveys in Britain has shown that most people

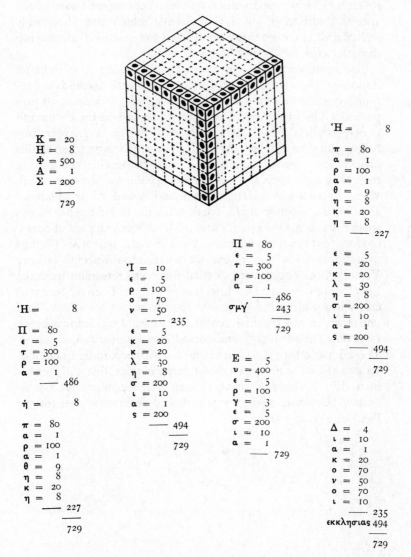

$K = 20$
$H = 8$
$\Phi = 500$
$A = 1$
$\Sigma = 200$
——
729

'H =   8

$\pi = 80$
$\alpha = 1$
$\rho = 100$
$\alpha = 1$
$\theta = 9$
$\eta = 8$
$\kappa = 20$
$\eta = 8$
—— 227

$\epsilon = 5$
$\kappa = 20$
$\kappa = 20$
$\lambda = 30$
$\eta = 8$
$\sigma = 200$
$\iota = 10$
$\alpha = 1$
$\varsigma = 200$
—— 494

729

'I = 10
$\epsilon = 5$
$\rho = 100$
$o = 70$
$\nu = 50$
---- 235
$\epsilon = 5$
$\kappa = 20$
$\kappa = 20$
$\lambda = 30$
$\eta = 8$
$\sigma = 200$
$\iota = 10$
$\alpha = 1$
$\varsigma = 200$
—— 494

729

$\Pi = 80$
$\epsilon = 5$
$\tau = 300$
$\rho = 100$
$\alpha = 1$
—— 486
σμγ΄   243
——
729

'H =   8

$\Pi = 80$
$\epsilon = 5$
$\tau = 300$
$\rho = 100$
$\alpha = 1$
—— 486

$\acute{\eta} =$ 8

$\pi = 80$
$\alpha = 1$
$\rho = 100$
$\alpha = 1$
$\theta = 9$
$\eta = 8$
$\kappa = 20$
$\eta = 8$
—— 227
——
729

$E = 5$
$\nu = 400$
$\epsilon = 5$
$\rho = 100$
$\gamma = 3$
$\epsilon = 5$
$\sigma = 200$
$\iota = 10$
$\alpha = 1$
—— 729

$\Delta = 4$
$\iota = 10$
$\alpha = 1$
$\kappa = 20$
$o = 70$
$\nu = 50$
$o = 70$
$\iota = 10$
——· 235
εκκλησιας 494
——
729

FIGURE 4 The stone of Cephas

regard 7 as lucky. We still speak of our 'seventh heaven' as a place of delight and remember the old belief that the seventh child of a seventh child has supernatural powers. Yet few are aware of the ancient tradition of the seven planets, which almost certainly started it all, nor even that God 'blessed and sanctified' the seventh day after creating the world in six.

The persistence of 40 is even more interesting. It links an important event in the Christian calendar with a common precaution against sickness. Both carry overtones of waiting or preparation. The forty days of Lent are a preparation for Easter and closely parallel the experience of the Babylonians. Every year, they had to wait through the forty days of the rainy season when clouds obscured the constellation of the Pleiades before they could enjoy their New Year festival which came when the wet weather ended. When 40 passed into Hebrew myth and legend it carried similar associations. Noah waited forty days for the Flood to end. Moses spent forty days fasting on Sinai and Jesus was 'tempted of Satan' for forty days in the wilderness. For a last example, which brings us straight back to the present, we can turn to fourteenth-century Venice whose doges were worried that ships returning from the Levant might bring a new infection of plague. To make sure that crews had a clean bill of health, they were not allowed to land until forty days after their arrival. This period was known as the *quarantina* ('about forty') and passed into English as quarantine. Though periods of isolation are now varied according to circumstances, the word gives us a direct numerological link with Jesus's forty days in the wilderness, as well as those miserable days in ancient Babylonia when the rain-soaked peasants waited for the Pleiades to reappear.[21]

10

Archetypes of Order

'I believe there are 15,747,724,136,275,002,577,605,653,
961,181,555,468,044,717,914,527,116,709,366,231,425,076,
185,631,031,296 protons in the universe and the same number
of electrons.'

Thus stated the distinguished astronomer, physicist, and
mathematician Sir Arthur Eddington in what, for me at least, is
the most arresting opening sentence of any chapter in scientific
literature. Sir Arthur admits that electrons can never be counted
because they cannot be distinguished from each other and are
never in a definite place. Yet he concludes:

'I am, however, strongly convinced that, if I have got the
number wrong, it is just a silly mistake, which would speedily
be corrected if there were more workers in the field. In
short, to know the exact number of particles in the universe
is a perfectly legitimate aspiration of the physicist.'[1]

Sir Arthur went even further. He believed that the relationships between the structure of different sub-atomic particles,
between the different components of the structure of the universe
and between sub-atomic articles and the universe as a whole could
be expressed in a series of ratios that never changed. Rom Harré,
the distinguished philosopher of science, expressed them like
this:

'1 the ratio of the mass of the proton to
   the mass of the electron,   value 1840
 2 the ratio of particle action to radiation
   action,   value 137
 3 the ratio of electrical to gravitational
   attraction,   value $2 \cdot 3 \times 10^{39}$

4 the ratio of the radius of curvature of
space-time to the mean value of the
wave (Schrödinger) for electron and
proton.   value $1 \cdot 2 \times 10^{39}$.'[2]

I do not pretend to understand the reasoning behind Sir
Arthur's theories and I am in good company. His attempt to find
all the numerical ratios ruling the world was never completed and
the best account we have is a posthumously published work
called *Fundamental Theory* (1946) edited by Sir Edmund Taylor
Whittaker. It is, says Professor A. Vibert Douglas, who wrote a
life of Eddington and is himself an astronomer, 'a book that is
incomprehensible to most readers and perplexing in many places
to all but which represents a continuing challenge to some.'[3]

For us, it is interesting as evidence that the ideas of Plato and
Pythagoras continue to thrive after more than three centuries of
experimental science. The odds against their survival once seemed
overwhelming, for after Copernicus, a new spirit of enquiry
gusted through the intellectual world, bringing men of learning
face-to-face with a dilemma as fearful as that which faced Victorian
traditionalists when Charles Darwin published *The Origin of
Species*. In the sixteenth and seventeenth centuries, the question
was not 'Did Adam have a navel?' but 'Was the harmony of the
spheres a jangling discord?' Most scientists felt forced to say 'yes'
though some contrived to keep a shred of doubt in some fireproof
corner of their minds. Those who said 'no' tended to be philosophers, theologians or mystics. One of the very few who managed
to straddle both worlds was Johannes Kepler (1571–1630), the son
of a German mercenary who became imperial mathematician to
the Holy Roman Emperor.

Kepler fully accepted the evidence of his senses, yet he was convinced that it would reveal a harmony that had existed in the mind
of God from all eternity. Since man's soul was an image of God,
the universal harmony was imprinted on that too and we could
recognize it by a sort of instinct. When observation and instinct
came together, true knowledge had been gained.

'For, to know is to compare that which is externally perceived with inner ideas and to judge that it agrees with them,
a process which Proclus expressed very beautifully by the

word "awakening", as from sleep. For, as the perceptible
things which appear in the outside world make us remember
what we knew before, so do sensory experiences, when consciously realized, call forth intellectual notions that were
already present inwardly; so that which formerly was hidden
in the soul, as under the veil of potentiality, now shines
therein in actuality.'[4]

He wrote too:

'The Christians know that the mathematical principles
according to which the corporeal world was to be created are
co-eternal with God . . . Geometry is co-eternal with the
Mind of God before the creation of things: it is *God Himself*
. . . [Geometry] has supplied God with the models for the
creation of the world. With the image of God it has passed
into man, and was certainly not received within through the
eyes.'[5]

A convinced Copernican, Kepler was nevertheless forced to
acknowledge the messiness of Copernicus's system. For Copernicus had assumed that the planets must move round the sun in
perfect circles at constant speeds and to account for apparent
irregularities in their motions, he had been driven to adopt
modifications to his original theory that made it intolerably complicated. Besides, at the imperial observatory near Prague, Kepler
had access to easily the most accurate and comprehensive series of
observations yet made, those of the team led by his predecessor
Tycho Brahe. Kepler personally studied those for Mars and
found that they could not be explained by Copernicus's theory,
even in its most convoluted form.

Now Kepler regarded the planets as material objects and hence
as inferior copies of the perfect ideas in God's mind. It seemed
to him, therefore, that they need not have the perfect circular
orbits prescribed by Copernicus and cast around for some other
geometrical figure that would fit the data. After years of work, he
found the solution, an ellipse with the sun at one of its two
focuses. Elliptical orbits fitted the data for the other planets too.
This became known as Kepler's First Law.

He now moved on to the next problem. It was obvious that the

planets did not travel at a constant speed along their elliptical
paths but again, he believed as an article of faith that they must
demonstrate some simple law. In a flash of intuition, he realized
that he had known the answer since his student days. The universe,
he had always believed, mirrored the Holy Trinity. The whole was
a sphere. The sun at its centre symbolized the Father; the inner
surface of the sphere marked by the fixed stars symbolized the
Son; and the space between in which the planets revolved
symbolized the Holy Ghost. So his problem was solved. 'The
sun,' he wrote, 'distributes its motive force through the medium
in which the movable things exist, just as the Father creates
through the Holy Ghost or through the power of the Holy
Ghost.'[6]

The sun's force, then, drove the planets round their orbits. It
radiated in separate lines from the sun's centre. The further these
lines penetrated into space, therefore, the greater the distance
between them. It followed that more would strike a given planet
when it was near the sun, less when it was further away. (Figure 1.)
So a planet's speed was directly proportional to its distance from
the sun.

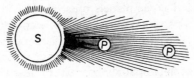

FIGURE 1 Effect of the sun's force on the same planet at
different distances

The theory was based on theology, not facts, but it fitted the
facts and in its refined form, Kepler's Second Law is still valid. It
states that as a planet moves round its ellipse, it sweeps out equal
segments in equal periods of time. Thus, in Figure 2, if the times
taken for a planet to move from P to P^1 and from P^2 to P^3 are
equal, so too are the areas PP^1S and P^2P^3S.

It was not until the appearance of his aptly entitled *Harmonies of
the World* in 1619 that Kepler stated his Third Law. Again, he
knew intuitively that there must be a simple mathematical ratio
relating the speeds of different planets but it took ten years' study

of the data to find the so-called 3/2 rule. This stated that the ratio between the squares of the time it took to complete their ellipses was equal to the ratio between the cubes of their average distances from the sun. This Third Law marked a completely new step in astronomy. It was the first time that any astronomer had defined not just the behaviour of a planet in its own orbit but the relationship between the speeds of different planets.

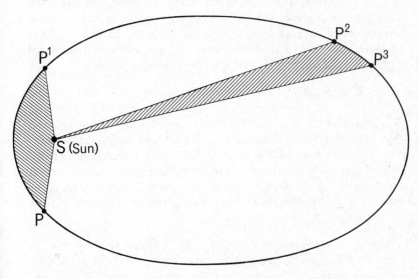

FIGURE 2 Kepler's Second Law: $PP^1S = P^2P^3S$

For us, though, the most interesting thing about Kepler is the way in which he discovered all three laws. He started with an idea of how things ought to be and could easily have gone on to work out a wholly speculative number system like so many of his predecessors. Instead, he matched his ideas against the scientific data and propounded his laws only when he found the correspondence he was expecting through a painstaking review of the evidence. Thomas S. Kuhn says of him:

'He was a mathematical Neoplatonist or Neopythagorean who believed that all of nature exemplified simple mathematical regularities which it was the scientist's task to discover. To Kepler and others of his turn of mind, a simple mathematical regularity was itself an explanation. To him, the Third Law in and of itself explained why the planetary orbits had been laid out by God in the particular way that they had, and that sort of explanation, derived from mathematical harmony, is what Kepler continually sought in the heavens.'[7]

Since Kepler, I doubt if any major scientist has admitted using a theological model as a working hypothesis but in recent years many have confessed to a belief in the harmony of the spheres. In 1919, Arnold Sommerfield, professor of theoretical physics at Munich. wrote of the new quantum physics:

> 'What we hear today in the language of the spectra is a real music of the spheres of the atom, a concord of whole numbers, numerical relations, a progressive order and harmony of all diversities . . . In the final analysis, all whole-number laws of the spectral line and the atom are derived from quantum theory. It is the mysterious organ on which Nature plays the music of the spectra, and according to whose rhythm it controls the structure of the atom and the nucleus.'[8]

Werner Heisenberg, author of the Uncertainty Principle in microphysics and a pupil of Sommerfeld's, also held a Platonic view as a result of his work on elementary particles. All natural phenomena, he believed, revealed the existence of a universal harmony existing as an independent idea.

Perhaps the most interesting suggestion of recent years was made by Wolfgang Pauli, the Austrian physicist who held chairs at both Zurich and Princeton and in 1945 won the Nobel Prize for Physics. He linked the numerical Platonism of the physicists with the findings of modern depth psychology. He was well qualified to make the connection, for his 'exclusion principle' had described the behaviour of certain subatomic particles in the periodic table of elements (see chapter 8) and he had also given considerable thought to the way in which scientists developed new ideas. Experimental evidence was not enough in itself to spark off a new theory, still less was mere logic. Intuition was usually far more important than either. Often, it seemed as though some unknown force was turning the scientist's attention in the right direction.

Pauli came to believe that there was 'a cosmic order independent of our choice and distinct from the world of phenomena'. He went on:

'Every partial recognition of this order in nature leads to the formulation of statements that, on the one hand, concern the world of phenomena and, on the other, transcend it by employing, "idealizingly", general logical concepts. The process of understanding nature as well as the happiness that man feels in understanding, that is, in the conscious realization of new knowledge, seems thus to be based on a correspondence, a "matching" of inner images pre-existent in the human psyche with external objects and their behaviour. This interpretation of scientific knowledge, of course, goes back to Plato . . .'

He was well acquainted with the work of Kepler who had described the images of which his intuition gave a glimpse as 'archetypal'.[9]

Kepler's use of the word 'archetypal' was interesting. For the psychologist C. G. Jung had used the term 'archetypes' to describe patterns of behaviour which all men carried in their unconscious minds. They held a strong emotional charge and were symbolized in images common to the art, myth, and religion of almost every culture. Examples are the Hero, the Great Mother, and the Tree of Life. Jung wrote:

'Although associated with causal processes, or "carried" by them, they continually go beyond their frame of reference . . . because the archetypes are not found exclusively in the psychic sphere but can occur just as much in circumstances that are not psychic (equivalence of an outward physical process with a psychic one).'[10]

Pauli noted the close similarity between Jung's archetypes and Kepler's primary images, both in themselves and in the way they worked as 'instincts of the imagination'. He continued:

'As ordering operators and image-formers in this world of symbolic images, the archetypes thus function as the sought-for bridge between the sense perceptions and the ideas and are, accordingly, a necessary presupposition even for evolving a scientific theory of nature.'[11]

Even more interesting, Jung held that number itself could be defined psychologically as '*an archetype of order* which has become conscious' and natural numbers therefore had 'an archetypal character'. He took the view that they were 'as much found as invented'; in other words they had always existed in their own right and were all along waiting to be discovered. Man 'invented' them only in so far as he became aware of their existence.[12]

Moreover, some numbers were not just archetypes of order. They took on archetypal meanings of their own. The number 4, for instance, has such deep roots in human experience that it may well be seen to express some quality far more profound than that shared by a collection of items greater than three but less than five. Apart from Empedocles' four elements, we have the tetragrammaton, the tetraktys, the four seasons, and the four points of the compass. Nearly always there is an implication of wholeness or completion, usually in a positive sense but occasionally negative. It is said that Charles Manson, the Californian ritual murderer, revered the Beatles pop group because he associated them with the four angels of Revelation 'to whom it was given to hurt the earth and the sea'. (Rev. 7.2.)

Most especially, the number four is the basis of the mandala, an ancient Eastern religious symbol commonly in the form of a circle containing a square or divided into quarters or multiples of four (Figure 3). Essentially, it is an idealized plan of the cosmos protected from disintegrating forces by the outer circle. It is used as an object of contemplation so as to help the mystic achieve a state of wholeness, reflecting in himself the fourfold unity of the cosmos. Sacred designs based on 4 and its multiples can be found among North American Indians, in Babylonian ziggurats, in all manner of crosses and in mediaeval Christian art where Christ is often shown at the centre of a square with the four evangelists at the corners. It occurs too in the four-square altar of Exodus (27.1), in the four-square court of Ezekiel (40.47) and in the four-square city of the New Jerusalem (Revelation 21.16).

Jung was especially interested in the quaternity because he found similar patterns in the therapeutic paintings of patients. Their dreams too might include a courtyard or a clockface, a flower or an eight-spoked wheel, a square room or four people in a boat.[13] Usually, these symbols emerged gradually during the

FIGURE 3 A typical mandala

course of therapy, becoming steadily more complete, so that the appearance of a well-developed mandala usually corresponded with a high degree of psychic integration.

There is further evidence of the archetypal power of numbers in a special form of magic square. The Chinese Empress Wu knew of it more than 1300 years ago when she built a huge

temple called the *Ming T'ang* or Hall of Light. It was dedicated
to heaven and had nine numbered rooms arranged in the form of
a square. Except for room 5, which was in the middle, each odd-
numbered room had a single dais and each even-numbered room
a double dais, making twelve in all, one for each month of the
year.

The Chinese year varied in length and citizens found it hard
to keep their diaries in order. At the beginning of each month, an
official called the Son of Heaven mounted the appropriate dais
and read out the Proclamation of Space and Time so that everyone
could be clear about the date. What gave the proclamation its
special authority was the arrangement of the rooms. For they
were not numbered in order but so as to form a magic square of
great potency (Figure 4). According to E. J. Holmyard: '. . . the
Son of Heaven, when in the *Ming-Tang*, was believed to become
the incarnation of deity and therefore possessed of unlimited
power over matter.'[14]

No one knows where the square originated but it was already
old in the seventh-century A.D. In China, it was called the *Lo-shu*
or pattern of the river Lo and seems to have been one of two
legendary number symbols that were later incorporated in the
cosmology of Taoism. It was, says Marie-Louise von Franz,

'a cosmic plan which, according to legend, was said to
have been given to the great culture hero Yü by a dragon-
horse or a god in the form of a tortoise. It contains a ground
plan of the universe . . .'[15]

FIGURE 4 Ming T'ang magic square

The symbol with which it was often associated was a number
cross known as Ho-t'u and cross and square were often regarded

as complements of each other. Elementary rules of arithmetic and geometry were derived from them, and together they formed the mathematical basis of the ancient Chinese number oracle *I Ching*. The fact that they both have 5 at the centre is significant because in ancient China, 5 symbolized earth as the great feminine and centering principle. Also, it was the middle term of the odd series of digits 1, 3, 5, 7, 9 and added to 6, the middle term of the even series 2, 4, 6, 8, 10, became 11, which is the number of the Tao or Way.

What makes the Lo-shu square especially interesting is its ubiquity. We find it venerated by civilizations of almost every period and continent. The Mayas of Central America held it in high esteem and today, it is used by the Hausa people of north-western Nigeria as a calculating device with magical associations. It was revered by the ancient Babylonians and was used as a cosmic symbol in prehistoric cave-scratchings in northern France.

It also had a special significance for the Muslims. In Islam generally it was thought to symbolize the power of Allah spreading round the earth and returning to its source. Members of secret societies used it as a code frame, linking the squares containing particular numbers by straight lines which were formed into a symbol. This was sent to fellow members who discovered the meaning, usually esoteric, by re-applying it to the square. A well-known example is the symbol as shown in Figure 5. Applied to the square, it crosses all the numbers except 8 (Figure 6). Idries Shah interprets the meaning as follows:

'Eight symbolizes the number of perfect expression, the octagon, representing, among other things, the cube. The figure also covers eight of a total of nine squares. The meaning here is "The eight (balance) is the way to the nine". Nine stands in Arabic for the letter *Ta*, whose hidden meaning is "secret knowledge".'[16]

In passing, it is worth noting that once again, multiples and powers of 4 are linked with perfection, completion, and balance. Also, the symbol is sometimes found as a mason's mark in mediaeval buildings, suggesting a link between secret societies and practising masons.

Geber, the eighth-century Islamic alchemist (or a later writer using his name) found in the same square a key to the elements. If it was divided in the manner shown in Figure 7, the four numbers in the bottom left-hand corner added up to 17, which

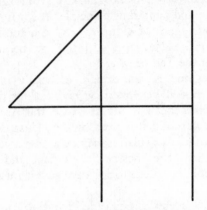

FIGURE 5 Islamic symbol

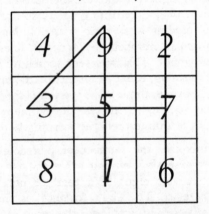

FIGURE 6 The symbol applied to the square

was regarded as highly auspicious in many cultures. For Christians, 17 symbolized the star of the Magi; for Kabbalists, the 17th path led to the reward of the righteous. Geber saw too that the remaining numbers of the square added up to 28, the number of letters in the Arabic alphabet.

Like other practical alchemists, Geber's aim was to transmute base metals into gold, an endeavour in which he used elixirs ranging from love-in-a-mist to the urine of a gazelle. First, however, he had to work out the composition of each base metal in terms of the four elemental qualities—hot and cold, wet and dry. He did this by constructing a table in which the four qualities

4	*9*	*2*
3	*5*	*7*
8	*1*	*6*

FIGURE 7 Geber's square

were each divided into 7 grades, giving 28 in all. The Arabic name for each metal happened to contain four consonants and by numbering these and relating them to his table, he was able, by an extremely complicated formula, to determine the metal's external constitution, By an even more complicated process, using the number 17, he converted this into the metal's internal constitution and was thus able to tell what changes were needed to change it into gold.

There is no evidence that Geber or any of his followers succeeded in their aim but the interesting fact is that they believed they could do so through the power of the Lo-shu square. These Arabian alchemists, together with the Empress Wu, Babylonian astrologer-priests, the prehistoric cavemen of France, the ancient Mayans of Yucatan, and modern Hausa tribesmen were all convinced in their different ways that the same pattern of numbers held the secret of the universe. It may be that some or all of them learned of the square from a single primal source. It may be that some or all of them stumbled across it independently. What is significant is that all believed that the square had supernatural attributes. If it had no power in itself (though many believed it had) all agreed that it reflected the power of the cosmos in terms of number. In the words of Marie-Louise von Franz, we may well come to the conclusion: 'It is not what we can *do* with numbers but what *they* do to our consciousness that is essential.'[17]

The theory of archetypes may also help to explain some of the extraordinary coincidences that occur in everyday life. Now it is only too easy to fall into the trap of finding special significance in every coincidence, for most of them can easily be explained by the laws of chance. If anyone doubts this, he should spend a few hours playing snap, a card game based solely on the coincidence of similar cards.

Some coincidences, though, are different. Something more than chance seems to be at work. A hint of purpose peeps through, as in an extraordinary case reported by Arthur Koestler in London's *Sunday Times*. Readers had been asked to send in accounts of co-incidences in their own lives and among them was a London policeman. In 1967, he said, a friend had repeatedly tried to ring him at the police station to which he was attached but always got the 'number unobtainable' signal. The reason was simple. The number had been changed and when they next met, the policeman told his friend that the new number was 40166. Next day, the policeman realized he had made a mistake. The new number was not 40166 but 40116. Before he managed to tell his friend of his error, he happened to be patrolling an industrial estate with a colleague. They saw a light in a factory and went to investigate. A telephone rang. The policeman picked it up. Astonished, he found that the caller was the friend to whom he had given the wrong number. He glanced down but there was no number on the instrument. When he contacted the factory manager, he found out that it was ex-directory and had been removed. The policeman asked what it was. The manager replied, 'It's 40166'. The policeman's friend had decided to ring him at the very moment when he was within hearing distance of the telephone bearing the number which he had been wrongly given.[18]

Coincidences of this type are examples of what Jung called 'synchronicity'. He defined synchronicity as 'an acausal connecting principle' linking a psychic process (the friend's wish to speak to the policeman) with an objective happening taking place at the same time (the proximity of the policeman to the phone bearing the wrong number). He believed that synchronicity might explain cases of clairvoyance, precognition, and intuitive methods of divination such as that based on the *I Ching*.

The question immediately arises as to how we can fit synchronicity into the accepted scientific view of the world. Surprisingly, it is not difficult. The principle of causality on which most science is based is sometimes true only statistically. For instance, the motions of sub-atomic particles do not individually obey the laws of cause and effect. All we can talk about are possibilities and probabilities. Once we dispense with causality, the way is open for some other explanation of how one event can be connected with another and synchronicity is a possibility that we must take into account. It works through the archetypes, including numbers. These make up an underlying pattern of order to which events correspond sometimes in terms of cause and effect, sometimes not. This may seem strange to us but the idea of correspondence was an assumption shared by most philosophers up to and including Leibnitz and the notion that cause-and-effect is the only way in which events may be connected can well be looked on as an aberration that has lasted a mere three hundred years and is no longer valid.

Jung wrote:

> 'Synchronicity is no more baffling or mysterious than the discontinuities of physics. It is only the ingrained belief in the sovereign power of causality that creates intellectual difficulties and makes it appear unthinkable that causeless events exist or ever could occur. But if they do, then we must regard them as *creative acts*, as the continuous creation of a pattern that exists from all eternity, repeats itself sporadically, and is not derivable from any known antecedents.'[19]

Jung warns, though, against the danger of thinking that an event has no cause simply because we have not yet found one. We may regard it as causeless 'only when a cause is not even thinkable.'[20]

There is a rather different kind of coincidence which does not seem to fit Jung's description of synchronicity but which also seems to call for some explanation other than chance. We find it in the fate of American presidents elected at intervals of twenty years since 1840. Without exception, they have died in office. They are:

> William Henry Harrison. Elected 1840. Died 1841.
> Abraham Lincoln. Elected 1860. Assassinated 1865.
> James A. Garfield. Elected 1880. Assassinated 1881.
> William McKinley. Elected 1900. Assassinated 1901.
> Warren G. Harding. Elected 1920. Died 1923.
> Franklin D. Roosevelt. Elected 1940. Died 1945.
> John F. Kennedy. Elected 1960. Assassinated 1963.

No one has ever suggested a good reason why the 'twenty-year presidents' should be marked out for early death. They varied in age from 46 to 68 and they came from both major parties. In a sense all their deaths were political in that the strain of office broke their health or they were victims of politically motivated assassins. Yet the American presidency is not normally a dangerous occupation. Only one other president, Zachary Taylor, has ever died in office and no other president has ever been assassinated. Presidents who have lived well beyond the allotted span include Andrew Jackson, who died at the age of 78, Van Buren 80, Herbert Hoover 90, Harry S. Truman 88, and Dwight Eisenhower 78. Why then should the twenty-year presidents share such a bizarre fate?

One man who studied coincidences of this kind was Paul Kammerer (1880–1926), the Austrian scientist who is best remembered for his controversial work on the midwife toad. In 1919, he published *The Law of Series* which was based on a study of a hundred coincidences and on systematic observations of 'chance' events. At every opportunity, sitting on a park bench or riding home on a tram, Kammerer would make detailed observations on everyone who came and went, noting down their apparent age, height, hair colour, type of hat worn, whether they were carrying a parcel or an umbrella, and so on. After allowing for obvious outside influences such as weather and time of day, he found that particular characteristics tended to bunch together. Red-headed men, for instance, did not appear randomly throughout the period of observation but tended to come in clusters with long gaps in between. We all notice this tendency in everyday life.

Good or bad luck seems to come in runs. 'It was one of those days,' we say, or 'It never rains but it pours'. Kammerer proposed a theory to explain this tendency for things or events to cluster in space or occur at regular intervals in time. He called it 'the Law of Seriality'.

It is not an easy concept to grasp and Arthur Koestler, who dubbed Kammerer 'the Linnaeus of coincidence', explained it like this:

'The central idea is that, side by side with the causality of classical physics, there exists a second basic principle in the universe which tends towards unity; a force of attraction comparable to universal gravity. But while gravity acts on all mass without discrimination, this other universal force acts selectively to bring like together both in space and in time; it correlates by affinity, regardless of whether the likeness is one of substance, form or function or refers to symbols. The *modus operandi* of this form, the way it penetrates the trivia of everyday life, Kammerer confesses to be unable to explain because it operates *ex hypothesi* outside the known laws of causality . . .

'Series in time i.e. the recurrence of similar events, he interprets as manifestations of periodical or cyclic processes which propagate themselves like waves along the time-axis of the space-time continuum. We are, however, only aware of the crests of the waves; these enter into consciousness and are perceived as isolated coincidences, whereas the troughs remain unnoticed (this, of course, is the exact reversal of the sceptic's argument that out of the multitude of random events we pick out those few which we consider significant). The waves of recurrent events may be kept in motion either by causal or a-causal i.e. "serial" forces.'[21]

The movements of the planets are the result of causal forces, while acausal forces would account for the fates of the twenty-year presidents and, according to Kammerer, for the biorhythmic periods which we discussed in chapter 3. The action of the law of seriality, said Kammerer, was 'ubiquitous and continuous, in life, nature and cosmos. The law of seriality is the umbilical cord that connects thought, feeling, science and art with the womb of the universe which gave birth to them.'[22]

If the idea seems strange, it is worth recalling that Einstein found it 'by no means absurd'.[23] It would not have perturbed Wolfgang Pauli, Sir Arthur Eddington, Kepler, Plato or Pythagoras. In their different ways, all believed in cosmologies which depended on some principle transcending mere cause-and-effect, not because they were stupid, wilful or misguided but because it seemed to them that content was preceded by form and that form could often be expressed in number.

I I

Conclusion

FROM the farthest deserts of outer space to the innermost recesses of the atom, number chimes through the universe with the subtle modulations of a polyphonic Mass. The vibrations are taken up by bees and bats, by daises and pineapples, by sun, moon and stars, by palolo worms and potatoes, and by man as mythmaker, poet, architect, gambler, and scientist.

We respond to these harmonies intuitively, delighting in the curves of a Botticelli Venus or the geometry of Islamic ornamentation, often without being aware of the numerical relations which they embody. Our feelings are engaged, and never more so than when listening to music which the philosopher Leibnitz described as 'a hidden practice of the soul, which does not know that it is dealing with numbers.'[1]

Number can be just as satisfying intellectually. As we work out the intricacies of a magic square or discover some unsuspected pattern in the behaviour of a little-known insect, we experience a feeling of being at home in the world. We and everything around us are dancing to the same beat.

There is reason for thinking that the dance comes before the dancers, the dance-floor or the instruments of the musicians. It is the form that precedes content. The case was put most eloquently by Professor Werner Heisenberg in his 'Athens speech' delivered on the Hill of Pnyx in 1964. He started by contrasting two opposing views of reality, that of Plato and other idealists who believed that it lay in ideas or forms, including mathematical forms such as the cube and the tetrahedron, and that of Democritus and other atomists who took the view that matter could be broken down into little, hard, indivisible lumps, rather like the 'billiard ball' atoms of Newtonian physics, and these were ultimate reality.

Between the sixteenth and nineteenth centuries, technological success gave the atomists the upper hand and it was still thought that the atom was indivisible. When it was finally split into nucleus and electrons, it seemed that these were the ultimate constituents of matter, but then the nucleus itself was split into protons and neutrons and it turned out that these too could be split, provided sufficient force was applied. Was there such a thing, then, as an ultimate smallest unit or could each successive sub-unit be re-split, and so on to infinity? If so, what was ultimately 'real'?

An interesting paradox emerged. When elementary particles broke up, *the fragments were as big as the original particles*. Professor Heisenberg explained such an apparent impossibility like this:

'... the best description of these collision phenomena is not to say that the particles have been split, but to speak about the creation of particles out of energy, according to the laws of relativity. We can say that all particles are made of the same fundamental substance, which may be called energy or matter; and to formulate: the fundamental substance "energy" becomes "matter" by assuming the form of an elementary particle. In this way, the recent experimental results have taught us, that we can combine the two conflicting statements: "matter is infinitely divisible" and "there are smallest units of matter" without running into a real contradiction.'

He added:

'I think that modern physics has definitely decided in favour of Plato. In fact these smallest units of matter are not physical objects in the ordinary sense; they are forms, ideas which can be expressed unambiguously only in mathematical language.'

The central problem of modern physics was to find an all-embracing principle by which the underlying symmetries of nature could be expressed mathematically.

'This situation reminds us at once of the symmetrical bodies which Plato introduced to symbolize the fundamental structure of matter. His symmetries have not been the correct ones; but Plato was right in believing that ultimately, in the centre of nature, we find mathematical symmetries.'[2]

A view of this kind put forward by one of the most distinguished physicists of the twentieth century might confirm the suspicion of many that scientists are taking only a reductionist view, that they are impoverishing human experience by reducing it to a string of numbers.

This is certainly true of much rationalistic thought, a trend of thinking that dates back in its present form to Galileo who wrote: 'The Book of Nature is written in mathematical characters.' In itself, this statement need not demean the life of man but the way in which Galileo developed the thought most certainly did. As Roger Poole put it:

'Galileo *shears away* the entire world of sense-impressions, emotions and the qualities that make up our everyday world. Then, he *substitutes* a knowledge of the mathematical properties of the world for that complete, total human world that we knew before, and counts himself richer by the exchange . . . But what of the world as we *experience* it, the world in which "bodies" are perceived as coloured, resinous, gritty, smelly, dry, pleasant and so on?' (Poole's italics.)

Poole rightly points out that Galileo's contempt for 'secondary qualities' leads to 'a double retreat from the totality; first from the totality of lived experience, and then from the totality of knowledge as such.'[3]

One can sympathize with these sentiments and they are without doubt applicable to Galileo himself and to those who share his philosophy. But many modern scientists, including some of the very best, do not rationalize so naively. They simply accept that number is one way—*only* one way but sometimes the most appropriate—of expressing and experiencing the oneness at the heart of reality. They do not deny that other ways of expression and experience also exist but rightly, they leave them to those who deal in myths and symbols, images and metaphors, and since these make possible a wider approach, Heisenberg at least admits that their language may be more important than the scientist's.

So it's rather like a delicious cake. The recipe exists before the cake and defines it in terms of the ratios of its ingredients. But knowing the recipe does not take away our appreciation of the cake's taste, appearance, smell, texture, the love and art that went into its cooking or the infinity of associations which any cake (Proust's *madeleine!*) conjures up. On the other hand, as the sales of cookery books show, many people get deep pleasure from the contemplation of recipes without actually trying them out, just as mathematicians take pleasure in pure, unapplied mathematics.

We find, therefore, a new philosophy emerging, or rather a very old one re-emerging. Number permeates the cosmos as salt flavours the sea. We can enjoy it consciously through mathematics or unconsciously through art. We cannot ignore it. We can only try to live in tune with it. We can enrich our lives and understanding, and hope to achieve wholeness only by adding our own voices to the harmony of the spheres.

BIBLIOGRAPHICAL NOTES

1. LE CORBUSIER, *The Modulor* (London: Faber, 1954), p. 220

Introduction

1. FRISCH, K. von, *Animal Architecture* (tr. Lisbeth Gombrich) (London: Hutchinson, 1975), p. 87
2. ibid., pp. 91–2
3. ibid., pp. 88, 89
4. ibid., p. 85
5. Quoted by BERGAMINI, D., *Mathematics* (New York: Time, 1963), p. 18
6. *The Guiness Book of Records*, edited and compiled by N. and R. McWhirter (Enfield: Guiness Superlatives, 1975), pp. 81–2. *The Guiness Book of World Records* (New York: Sterling, 1975)
7. *Sunday Times*, London, 30 June 1974
8. *Sunday Times*, London, 24 November 1974

1. TO THE POWER OF TEN

1. *The Times*, London, 12 May 1975
2. DROSCHER, V. B., *The Magic of the Senses* (tr. U. Lehrburger and O. Coburn) (London: W. H. Allen, 1969), pp. 148–58. (New York: Dutton, 1969)
   See also BURTON, R. *Animal Senses* (Newton Abbot: David & Charles, 1970), pp. 62–70
3. DANTZIG, T. *Number, the Language of Science* (London: Allen and Unwin, 1930), p. 3. (New York: Macmillan, 1930)
4. LANYON, W. E. *Biology of Birds* (London: Nelson, 1964), p. 118. (Garden City, N.Y.: Natural History Press, 1964)
5. THORPE, W. H. *Animal Nature and Human Nature* (London: Methuen, 1974), pp. 299–300. (Garden City, N.J.: Anchor Press, 1974)
6. WOODWORTH, R. S. and SCHLOSBERG, H. *Experimental Psychology*, third edition revised (London: Methuen, 1954), p. 94. (New York: Holt, 1954)

7. STRUICK, D. J. *A Concise History of Mathematics* (London: Bell, 1962), p. 1. (New York: Dover, 1948)
8. DANTZIG, op. cit., p. 7
9. ibid., p. 11
10. HERODOTUS, *History* (tr. G. Rawlinson. ed. M. Komroff) (New York: Tudor, 1947), p. 377
11. STRUICK, op. cit., p. 3
12. Quoted by DANTZIG, op. cit., p. 15
13. STRUICK, op. cit., p. 4
14. DANTZIG, op. cit., p. 13
15. ORE, O. *Number Theory and Its History* (New York: McGraw-Hill, 1948), p. 2
16. ibid., p. 2
17. ibid., p. 38

2. WHAT TIME IS OUR BODY?

1. WHITE, G. *The Natural History of Selborne*, Everyman's Library (London: Dent, 1906), p. 136. (New York: Dutton, 1906)
2. KERNER von Marilaun, A. *The National History of Plants* (London: Blackie, 1895), vol. 2, pp. 215–16
3. CARSON, R. L. *The Sea Around Us* (London: Staples, 1951), p. 163. (New York: Oxford University Press, 1951)
4. ibid., pp. 165–6
5. GAUQUELIN, M. *The Cosmic Clocks* (London: Owen, 1969), p. 103. (Chicago: Regnery, 1967)
6. ibid., pp. 106–8
7. LUCE, C. G. *Body Time* (London: Temple Smith, 1972), p. 22. (New York: Pantheon, 1971)
8. ibid., p. 21
9. CONROY, R. T. L. W. and MILLS, J. N. *Human Circadian Rhythms* (London: Churchill, 1970), p. 10
10. LUCE, op. cit., p. 159
11. ibid., pp. 41–2
12. CONROY and MILLS, op. cit., p. 104
13. ibid., p. 22
14. LUCE, op. cit., p. 44
15. ibid., p. 43
16. *Daily Mail*, London, 31 December 1970
17. *Daily Mail*, London, 24 November 1971
18. LAVIE, P. and KRIPKE, D. 'Who's got rhythm?' *Psychology Today* (British Edition) 1975, vol. 1, No. 8, pp. 31–3

19. GAUQUELIN, op. cit., pp. 154–6
20. ibid., pp. 157–8
21. ibid., pp. 159–60

3. ARE OUR LIVES RULED BY NUMBERS?

1. Information supplied by the Biorthythmic Research Association (B.R.A.), Normanton-on-Soar, Loughborough, Leicestershire
2. Quoted in FREUD, S. *The Origins of Psychoanalysis. Letters to Wilhelm Fliess. Drafts and Notes: 1887–1902* (London: Imago, 1954), p. 7. (New York: Basic Books, 1954)
3. JONES, E. *Sigmund Freud* (London: Hogarth Press, 1953), vol. 1, pp. 210 and 328–34. (New York: Basic Books, 1953)
4. ibid., pp. 320, 349
5. B.R.A. as above.
6. *idem*
7. *The Times*, London, 10 May 1973
8. B.R.A. as above
9. *idem*
10. ANDERSON, R. K., 'Biorhythm—Man's Timing Mechanism', *Journal of the American Society of Safety Engineers*, February 1963, pp. 17–21
11. BERKOVITCH, I. 'We Have the Rhythm . . . But Does it Affect the Risk of Accidents?' *Protection*, 1974, vol. 11, pp. 8–9
12. JOLLES, K. E. 'Biorhythms', *Nursing Times*, 15 February 1973, pp. 206–8
13. B.R.A. as above
14. JOLLES, K. E. Letter in *Motor*, 23 February 1974
15. Letter to the author. 25 November 1974
16. AHLGREN, A. 'Biorhythms', *International Journal of Chronobiology*, 1974, vol. 2, pp. 107–9

4. WHY IS 13 UNLUCKY?

1. Reported in *Psychology Today*, British edition 1975, vol. 1, No. 1, p. 10
2. *Evening News*, London, 14 April 1970
3. *Daily Express*, London, 16 August 1963
4. BARDENS, D. *Princess Margaret* (London: Hale, 1964), p. 19. (New York: Abelard Schuman, 1965)
5. JAHODA, G. *The Psychology of Superstition* (London: Allen Lane, 1969), p. 8

220   *Numberpower*

6. *Encyclopaedia Britannica* (Chicago: 1976) Marcopaedia, vol. 3, p. 598
7. FRAZER, J. *The Golden Bough*, abridged edition (London: Macmillan, 1922), p. 359. (New York: Macmillan, 1923)
8. HUGHES, P. *Witchcraft* (London: Longmans, 1952), p. 104. (Baltimore: Penguin, 1965)
9. OPIE, P. and I. *The Lore and Language of Schoolchildren* (Oxford: Clarendon Press, 1959), p. 208
10. GRAVES, R. *The White Goddess* (London: Faber, 1959), p. 185. (New York: Vintage, 1958)
11. GAUQUELIN, M. *Astrology and Science* (London: Peter Davies, 1970), p. 192. *The Scientific Basis of Astrology* (U.S. Title) (New York: Stein and Day, 1969)
12. HUFF, D. *Cycles in Your Life* (London: Gollancz, 1965), p. 112. (New York: Norton, 1964), p. 112
13. ibid., pp. 115–16
14. ibid., p. 116
15. ibid., p. 116
16. FRAZER, op. cit., p. 606
17. (New York: Emerson, 1938) quoted in LANGSTAFF, J. *Adam's Rib* (London: Allen and Unwin, 1954), p. 47
18. LUCE, G. G. *Body Time* (London: Temple Smith, 1972), pp. 201–2. (New York: Pantheon, 1971)
19. DALTON, K. *The Menstrual Cycle* (Harmondsworth: Penguin, 1969), pp. 120–34

5. ODDS, CHANCE, AND LUCK

1. CONRAD, J. *The End of the Tether*, 'Youth: A Narrative and Two Other Stories' (Edinburgh and London: Blackwood, 1902), pp. 293–5
2. WYKES, A. *Gambling* (London: Aldus, 1964), p. 227
3. *Daily Mail*, London, 1 January 1973
4. LEVINSON, H. C. *The Science of Chance* (London: Faber, 1952), p. 20. (New York: Rinehart, 1952)
5. RICKMAN, E. *Come Racing With Me* (London: Chatto, 1952), pp. 153–4
6. WYKES, op. cit., p. 206
7. ibid., p. 227
8. MARRYAT, Captain F. *Peter Simple* Everyman edition (London: Dent, 1907), p. 43. (New York: Dutton, 1907)
9. LEVINSON, op. cit., p. 65

10. WYKES, op. cit., p. 46
11. LEVINSIN, op. cit., p. 68
12. WYKES, op. cit., pp. 226–7
13. ARNOLD, P. *The Book of Gambling* (London and New York: Hamlyn, 1974), p. 57
14. *Daily Mail*, London, 18 December 1973
15. See MACKENZIE, A. *Riddle of the Future* (London: Barker, 1974)
16. THORP, E. O. *Beat the Dealer* (New York: Random House, 1966), p. 4
17. ibid., chapters 6, 7 and 8
18. ibid., chapter 5
19. ibid., chapter 9

6. LIES, DAMNED LIES, AND STATISTICS

1. GARDNER, M. *Mathematical Puzzles and Diversions* (London: Bell, 1961), p. 44. (New York: Simon and Schuster, 1959)
2. MORONEY, M. J. *Facts from Figures* (Harmondsworth and Baltimore: Penguin, 1951), pp. 8–9
3. GARDNER, op. cit., pp. 41–3
4. ibid., p. 43
5. LEVINSON, H. C. *The Science of Chance* (London: Faber, 1952), pp. 248–9. (New York: Rinehart, 1952)
6. DAVIDSON, J. H. *Offensive Marketing* (London: Cassell, 1972), pp. 186–7
7. TALLENTIRE, D. R. 'The Mathematics of Style', *The Times Literary Supplement*, London, 13 August 1971
8. HEASMAN, M. A. and LIPWORTH, L. *Accuracy of Certification of Cause of Death* (London: H.M.S.O., 1966), pp. 51, 53
9. LEVINSON, op. cit., p. 233
10. HUFF, D. *How to Lie with Statistics* (London: Gollancz, 1962), p. 60. (New York: Norton, 1954)
11. *The Times*, London, 17 March 1976
12. *Investors Chronicle*, London, 2 April 1976
13. *Daily Mail*, London, 27 January 1976
14. *Daily Mail*, London, 29 January 1976
15. DEWEY, E. R. (with MANDINO, O.) *Cycles* (New York: Hawthorn, 1971), p. 108
16. ibid., p. 95
17. *The Times*, London, 9 April 1975
18. Quoted in DEWEY, op. cit., p. 96
19. *The Times*, London, 6 January 1975

20. *Sunday Times*, London, 3 November 1974
21. DEWEY, op. cit., p. 136
22. *Daily Mail*, London, 28 April 1969
23. *The Times*, London, 14 February 1975
24. ELLIS, K. *Prediction and Prophecy* (London: Wayland, 1973), p. 139
25. DEWEY, op. cit., pp. 143–4
26. *The Times*, London, 1 November 1975
27. *Encyclopaedia Britannica* (Chicago: 1976) Macropaedia, vol. 14, pp. 562–3
28. *Sunday Times*, London, 3 November 1974
29. ibid.
30. *Daily Mail*, London, 17 May 1969
31. *Sunday Times*, 16 May 1976
32. ibid.
33. *The Times*, London, 18 March 1976
34. *Sunday Times*, London, 3 November 1974
35. LANGLEY, R. *Practical Statistics* (London: Pan, 1968), p. 19, quoting COHEN, M. and NAGEL, E. *An Introduction to Logic* (London: Routledge, 1963). U.S.: LANGLEY (New York: Drake, 1971); COHEN and NAGEL (New York: Harcourt Brace, 1962)
36. ibid., p. 39, quoting SLONIM, M. J. *Sampling in a Nutshell* (New York: Simon and Schuster, 1960)
37. ibid., p. 38. See also LEVINSON, op. cit., pp. 236–8
38. LANGLEY, op. cit., p. 46
39. ibid., p. 149

7. THE FASCINATION OF NUMBERS

1. WATTS, A. *Beyond Theology* (London: Hodder, 1964), p. 37. (New York: Pantheon, 1964)
2. Quoted by ORE, O. *Number Theory and its History* (New York: McGraw-Hill, 1948), p. 180
3. ibid., p. 123
4. ibid., pp. 116–7
5. Quoted by GARDNER, M. *Mathematical Puzzles and Diversions* (London: Bell, 1961), pp. 131–2. (New York: Simon and Schuster, 1959), from *Current Literature*. vol. 2, April 1889, p. 349
6. REICHMANN, W. J. *The Fascination of Numbers* (London: Methuen, 1957), p. 130. (Fair Lawn, N.J.: Essential Books, 1957)
7. ORE, op. cit., p. 39

8. SIMON, W. *Mathematical Magic* (London: Allen and Unwin, 1965), pp. 25–8. (New York: Scribner, 1964)
9. REICHMANN, op. cit., p. 42
10a. REICHMANN, op. cit., pp. 147–8. *b.* SIMON, N., op. cit., pp. 111–14. *c.* SIMON, N., op. cit., pp. 115–17
11. BARLOW, P. E. *Theory of Numbers* (London: Johnson, 1811), p. 43
12. *The Guiness Book of Records* edited and compiled by N. and R. McWhirter (Enfield: Guiness Superlatives, 1975), p. 82. *The Guiness Book of World Records* (New York: Sterling, 1975)
13. Quoted by ORE, op. cit., p. 97
14. *The Times*, London, 12 March 1975
15. *The Times*, London, 27 July 1975

8. DIVINE PROPORTION

1. *Encyclopaedia Britannica* (Chicago: 1976) Macropaedia, vol. 10, p. 818
2. ibid., p. 818
3. DANTZIG, T. *Number, the Language of Science* (London: Allen and Unwin, 1930), p. 232. (New York: Macmillan, 1930)
4. BENTHALL, J. *Science and Technology in Art Today* (London: Thames and Hudson, 1972), p. 157–8. (New York: Praeger, 1972). See also JENNY, H. *Cymatics* (Basle: Brasiliano Presse, 1967)
5. ibid., pp. 62–3
6. ibid., pp. 114–5
7. Quoted by WARRY, J. G. *Greek Aesthetic Theory* (London: Methuen, 1962), p. 34. (New York: Barnes and Noble, 1970)
8. BUTLER, C. *Number Symbolism* (London: Routledge, 1970). pp, 168–9, 175. (New York: Barnes and Noble, 1970), quoted from REICH, W. *Alban Berg* (London: 1965), p. 143
9. NYMAN, M. *Experimental Music* (London: Studio Vista, 1974), pp. 51, 136
10. JORDAN, R. F. *A Concise History of Western Architecture* (London: Thames and Hudson, 1969), p. 170. (New York: Harcourt Brace and World, 1970)
11. Quoted by HONOUR, H. *The Companion Guide to Venice* (London: Collins, 1965), p. 107
12. Quoted by WITTKOWER, R. *Architectural Principles in the Age of Humanism* (London: The Warburg Institute, University of London, 1949), p. 20. (New York: Random House, 1965)
13. SCHOLFIELD, P. H. *The Theory of Proportion in Architecture* (Cambridge: University Press, 1958), pp. 21–6
14. Quoted by SCHOLFIELD, op. cit., p. 55

15. WITTKOWER, op. cit., p. 91
16. ibid., p. 41
17. ibid., pp. 40–4
17a. ibid., p. 96
18. ibid., p. 101
19. ibid., pp. 101–2
20. Quoted by SCHOLFIELD, op. cit., pp. 83 and 76
21. LE CORBUSIER, *The Modulor* (London: Faber, 1954), p. 27. (Cambridge: Harvard University Press, 1958)
22. ibid., p. 26
23. ibid., p. 37
24. ibid., p. 48
25. ibid., pp. 48–52
26. ibid., p. 56
27. LE CORBUSIER, *Modulor 2* (London: Faber, 1958), p. 44. (Cambridge: Harvard University Press, 1958)
28. SCHOLFIELD, op. cit., p. 124
29. LE CORBUSIER, *The Modulor*, pp. 62–3
30. FISCHER, E. *The Necessity of Art* (Harmondsworth: Penguin, 1963), p. 152. (Baltimore: Penguin, 1963)
31. ibid., p. 153
32. ibid., pp. 153–4
33. LE CORBUSIER, *Modulor 2*, p. 19
34. ibid., p. 156
35. READ, H. *The Philosophy of Modern Art* (London: Faber, 1964), p. 200. (Freeport, N.Y.: Books for Libraries Press, 1971)
36. ibid., p. 201
37. LE CORBUSIER, *The Modulor*, p. 219

9. IS GOD A NUMBER?

1. *The Times*, London, 24 April 1976
2. Quoted in DANTZIG, T. *Number, the Language of Science* (London: Allen and Unwin, 1930), p. 41. (New York: Macmillan, 1930)
3. PLATO, *Timaeus*. Loeb Classical Library (tr. R. G. Bury) (London: Heinemann, 1929), p. 59. (Cambridge: Harvard University Press, 1929)
4. ibid., pp. 79 and 73
5. See BUTLER, C. *Number Symbolism* (London: Routledge, 1970), p. 27. (New York: Barnes and Noble, 1970)
6. Quoted by BUTLER, op. cit., from St Augustine. *The City of God*

(tr. J. Healey), revised R. G. V. Tasker. Everyman's Library (London: Dent, 1960). (New York: Dutton, 1960)
7. BUTLER, op. cit.,
8. MATHEW, G. *Byzantine Aesthetics* (London: Murray, 1963), p. 26. (New York: Viking, 1964).
9. GRANT, R. M. (ed) *Gnosticism, An Anthology* (London: Collins, 1961), pp. 162–6
10. MANSEL, H. L. *Gnostic Heresies in the First and Second Centuries* (London: Murray, 1875), pp. 184–202
11. GRANT, R. M., op. cit., pp. 164 and 168
12. PONCÉ, C. *The Kabbalah* (London: Garnstone, 1974), p. 171
13. ibid., pp. 174–5
14. ibid., p. 176
15. See YATES, F. A. *Giordano Bruno and the Hermetic Tradition* (London: Routledge, 1964) (Chicago: University of Chicago Press, 1964)
16. BROWNE, T. *Urne Buriall and The Garden of Cyrus*, ed. J. Carter (Cambridge: University Press, 1958), pp. 109–10
17. BUTLER, op. cit., p. 133
18. FOWLER, A. *Spenser and the Numbers of Time* (London: Routledge, 1964), p. 3. (New York, Barnes and Noble, 1964)
19. ibid., p. 4
20. BOND, F. B. and LEA, T. S. *A Preliminary Investigation of the Cabala* (Oxford: Blackwell, 1917) end-paper
21. See BUTLER, op. cit., p. 28

10. ARCHETYPES OF ORDER

1. EDDINGTON, A. *The philosophy of Physical Science* (Cambridge: University Press, 1939), pp. 170–1. (New York: Macmillan, 1939)
2. HARRÉ, R. *The Anticipation of Nature* (London: Hutchinson, 1965), p. 87. (New York: Humanities Press, 1965)
3. *Encyclopaedia Britannica* (Chicago: 1976) Macropaedia, vol. 6, p. 297
4. Quoted in JUNG, C. G. and PAULI, W. *The Interpretation of Nature and the Psyche* (London: Routledge, 1955), p. 162. (New York: Pantheon Books, 1955)
5. ibid., Pauli, p. 164
6. ibid., Pauli, p. 173
7. KUHN, T. S. *The Copernican Revolution* (Cambridge: Harvard University Press, 1970), p. 217
8. Quoted in FRANZ, M–L. von, *Number and Time* (London: Rider, 1974), p. 46. (Evanston: Northwestern University Press, 1974)

9. Pauli in JUNG and PAULI, op. cit., p. 152
10. Jung in JUNG and PAULI, op. cit., p. 137
11. Pauli in JUNG and PAULI, op. cit., p. 153
12. Jung in JUNG and PAULI, op. cit., pp. 57–9
13. JUNG, C. G. *Psychology and Religion* (New Haven: Yale University Press, 1938), p. 65
14. HOLMYARD, E. J. *Alchemy* (Harmondsworth: Penguin, 1957), p. 39
15. von FRANZ, op. cit., p. 22
16. IDRIES SHAH, *The Sufis* (London: W. H. Allen, 1964), p. 191. (New York: Doubleday, 1964)
17. von FRANZ, op. cit., p. 33
18. *Sunday Times*, London, 5 May 1974
19. Jung in JUNG and PAULI, op. cit., p. 141
20. ibid., p. 142
21. KOESTLER, A. *The Case of the Midwife Toad* (London: Hutchinson, 1971), p. 140. (New York: Random House, 1972)
22. Quoted in KOESTLER, op. cit., p. 142
23. Quoted in KOESTLER, op. cit., p. 142

11. CONCLUSION

1. Quoted in FRANZ, M–L. von, *Number and Time* (London: Rider, 1974), p. 33. (Evanston: Northwestern University Press, 1974)
2. HEISENBREG, W. *Natural Law and the Structure of Matter* (London: Rebel Press, 1970), pp. 31, 32, 42
3. POOLE, R. *Towards Deep Subjectivity* (London: Allen Lane, 1972), p. 82

INDEX

Index

Note: Entries consisting of numbers are listed after the alphabetical section, and have been italicised to distinguish them from the page references.

Telephone-use volume, 108
Teleprinter operators, 25
Teltscher, Alfred, 40
Temperature cycles
  human, 21, 22, 25
  terrestrial, 113
Tennis scoring system, 9
Tennyson, Alfred, Lord, 17
Tetragrammaton, 184, 185
Tetraktys, 175–6, 183
Thommen, George, 46
Thorp, Edward O., 95, 97–8
Time
  effect of crossing zones on circadian rhythms, 28
  sexagesimal system, 9
Totalisator, 76
Tree alphabet, 64
Triangular numbers, 126
Triskaidekaphobia, *see 13*
Trissino, Giangiorgio, 155
Truman, Harry S., 117
Tsai, Wen-Ying, 150
Tynner, Julian, 91

Ultradian rhythms, 32–3
Uncertainty Principle, 200
United States Presidents
  death in office, xv, 210
  elections, 117–18
Universe, 178
Unlucky numbers, 55–70
Urine, 22

Vaccination results, 121
Valentinians, 182–3
Van Flandern, Thomas, xiii
Venice, 155
Vesica piscis, 152, 153
Vibration, 148
  of rods in strobe lighting, 150
  see also Acoustics
Vingt-et-un, see Blackjack
Vitruvius, 155–7, 161
Von Franz, Marie-Louise, 204, 208
Von Frisch, Karl, xii

Wagner, Richard, 151
Wasps, 3
Watts, Alan, 122
Weather
  influence of cycles of solar activity, 19–20

patterns, 111–16
  linked to odd- and even-numbered years, 112
  long-term military significance, 115
Wedding rites, 61
Wells, Charles Deville, 85
White, Gilbert, 17
Whittaker, Sir Edmund Taylor, 196
Witches
  broomsticks of elder, 64
  covens, 61
Worcestershire relics of moon worship, 60
Worker efficiency, 24–7
World War II, 190
Wrought iron price cycles, 109

Xerxes, 7–8

Yates, Francis, 185

1, 185
2, 185
3, 185
4 archetypal significance, 202
  use in psychotherapy, 203
5 conjugal number, 185–6
  in Chinese symbolism, 205
6 in Byzantine art, xiv, 181
  perfect number and length in days of Creation, 180
7 in St Ives puzzle, 129
  origin as 'lucky' number, 194
  relationship to cubed numbers, 129
9, 135
12, 183
13, xiv, 55–70
  patterns in ritual groupings, 60–1
  ploys to avoid its use, 55–6
17, 206
20, 10
21, 145
30, 183
34, 145
40, 194
221, 124
312, 124
142,857, 124
1089, 124